CAUSES OF PR
& POWER STRUGGLES
IN SOUTH SUDAN

GEN. STEPHEN BUAY ROLNYANG

The publisher wishes to acknowledge and thank Dr. Douglas H. Johnson for his invaluable help and support for Africa World Books and its mission of preserving and promoting African cultural and literary traditions and history. Dr. Johnson and fellow historians have been instrumental in ensuring that African people remain connected to their past and their identity. Africa World Books is proud to carry on this mission.

ISBN: 9780645819519

Cover design, typesetting and layout: Africa World Books
Unit 3, 57 Frobisher St, Osborne Park, WA 6017
P.O. Box 1106 Osborne Park, WA 6916

Table of Contents

About the Author

Born on November 15, 1968, in Mayom, Unity State Upper Nile Region, South Sudan, General Stephen Buay Rolnyang, started attending school at the age of nine years in Ajak-Kuac Payam, Twic County, Bhar El Ghazal region. Six years later, General Buay dropped out of school and joined the Anyanya-II Southern Sudan insurgency. In 1984, he abandoned the Anyanya-II camp and joined Sudan People Liberation Movement/Army (SPLM/A). After serving in the insurgency for one-year, General Paulino Matip Nhial, who had recruited him established Anyanya-II Movement in Western Upper Nile in 1985 and decided to release him whereby he returned to school.

General Buay proceeded to study in Khartoum, where he did his senior school certificate and thereafter rejoined General Paulino Matip Nhial in 1991. When he resumed work with the Anyanya-II movement, he was dispatched for special force training at the Sudanese Military Training Centre in Omdurman. After a year, General Buay went for specialized training with the Sudan armed forces' signal corps in Khartoum Bhari. On completion, General Paulino Matip Nhial deployed him as a radio operator at his headquarters. In 1992, General Buay joined the SPLA Nasir faction led by Riek Machar Teny Dhurgon and he was transferred in the same year to the headquarters of Riek Machar as a signal officer with the rank of a captain.

In 1996, General Buay was promoted to the rank of Alternate Commander (A/Cdr) and assigned as Chief of Signal in Western Upper Nile. However, when General Paulino Matip broke away from Riek Machar's faction, General Buay was promoted to the rank of full commander in the South Sudan United Movement/Army (SSUM/A) and assigned as Commander of Signal Corps. Nonetheless, General Buay defected from SSUM/A to the SPLM/A mainstream on August 10, 1998, where he was deployed as operations officer on the Twic border with Western Upper Nile, Mayom County under the direct

4

command of Salva Kiir Mayardit, who was by then the Commander of the SPLA 3rd Front.

In 2000, General Buay was redeployed as Chief of Logistics for the SPLA 3rd Front Bhar El Ghazal region under the overall command of Commander Salva Kiir Mayardit. He was sent for a one-year training course at the SPLA Institute for Strategic Studies at Lasu, Yei area on the South Sudan border with DRC Congo. Commander Salva Kiir Mayardit later sent General Buay to Western Upper Nile to assist Commander Peter Gatdet with the organizational structure of the SPLA when Peter Gatdet rejoined the SPLM/A in 1999.

From 2003 to 2006, General Buay was confirmed as the Commander of Western Upper Nile when Commander Peter Gatdet defected again to the Paulino Matip faction. From 2005 to 2007, General Buay was transferred from Western Upper Nile to Kapoeta, Eastern Equatoria where he commanded the SPLA 21st Infantry Mobile Independent Brigade. General Buay was later redeployed from Kapoeta to Juba SPLA General headquarters to establish the command of the SPLA military police and commando units, also known as special forces.

Between 2008 and 2014, he studied and obtained a Bachelor of Management Science (BMS) degree from Kyambogo University (Uganda). He also undertook several other certificate courses from Cambridge International College and the SPLA Senior Command and Staff Course (SCSC) conducted in Juba at the SPLA General Headquarters by American military experts. After completion of the courses, he was redeployed from the SPLA Special Forces' unit to the SPLA 1st Infantry Division in Upper Nile Region, with its headquarters in Renk in 2013, as deputy commander of the division, but shortly between 2014 and 2015, he was confirmed as Commander of the SPLA 1st Infantry Division when General Angelo Jongkuc - the then commander was redeployed to the SPLA General Headquarters due to the crisis the country underwent in December 2013.

From 2015 to 2017, General Buay was redeployed from the SPLA 1st Infantry Division to command the SPLA 4th Infantry Division in Western Upper Nile with its headquarters in Bentiu, but shortly thereafter, he was arrested and detained before being re-assigned as Director General of Procurement in the Ministry of Defence and Veterans Affairs after he had been released from jail.

Between 2017 and 2018, General Buay was redeployed for the second time to command the SPLA 4th Infantry Division in Unity State. He was then further redeployed to command the SPLA 5th Infantry Division in Western Bhar El Ghazal with its headquarters in Wau in 2018. He was redeployed from Wau 5th Infantry Division to the SPLA General headquarters as Director of the Military Organization on May 22, 2018, until he was arrested on May 31, 2018, on allegations of attempting to rebel against the government of South Sudan. However, the military tribunal, which heard the case, acquitted him of the charges levelled against him. After spending three years at home, General Buay fled the country in 2021, and joined the South Sudan United Front SSUF/A) led by General Paul Malong Awan, who promoted him to the rank of Lieutenant General. However, General Buay did not serve for long in the SSUF/A as he pulled out from the South Sudan United Front and formed his own movement called South Sudan People's Movement/Army (SSPM/A), and he became the chairman and Commander-in-Chief of the SSPM/A.

Dedication

This book is dedicated to my fellow South Sudanese political activists who have been subjected to unspeakable injustices, torture, and intimidation, languishing in the custody of South Sudan's notorious Blue-house for years without trial for crimes they never committed. I dedicate this book also to General Paulino Matip Nhial, the founding father of the Anyanya-II in Western Upper Nile and Bul territorial militias. Paulino Matip recruited thousands of Bul youth into the Anyanya-II and raised and nurtured sons and daughters of Bul from militia army officers to National Army Officers in the Republic of South Sudan.

Acknowledgment

I would like to express my gratitude to all my friends whom I will not mention for providing me with the necessary information, and to my family for moral support. Without their support, this book would not have become a reality.

Abbreviations

1st Lt	1st Lieutenant.
White Army	A name was given to a local militia in Upper Nile Region.
Anyanya	A name was given to a rebel movement in South Sudan.
J-1	The presidential palace in Juba.
Torit	A town in Eastern Equatoria.
Nasir	A town in Upper Nile.
AU	African Union.
ARCSS	Agreement on the Resolution of the Conflict in South Sudan.
A/Cdr	Alternate Commander.
APMHC	Alternate Political-Military High Command.
COGS	Chief of General staff.
Cdr	Commander.
CPA	Comprehensive Peace Agreement.
EU	European Union.
GCM	General Court Martial.
GOSS	Government of South Sudan.
HQ	Headquarters.
IGAD	Inter-Governmental Authority on Development.
NCP	National Congress Party.
NCF	National Consensus Forum.
NCS	National Consensus statement.
PMHC	Political-Military High Command.
R-ARCSS	Revitalized Agreement on the Resolution of the Conflict in South Sudan.
RTGONU	Revitalized Transitional Government of National Unity.
RTC	Roundtable Conference.

SSDF	South Sudan Defence Force.
SSIM/A	South Sudan Independence Movement/Army.
SSLM/A	South Sudan Liberation Movement/Army.
SSOA	South Sudan Opposition Alliance.
SSPDF	South Sudan People's Defence Forces.
SSUM/A	South Sudan United Movement/Army.
SF	Southern Front.
SPG-9	Soviet Anti-tank type gun.
SANU	Sudan African National Union.
SAF	Sudan Armed Forces.
SPDF	Sudan People Democratic Front.
SPLA	Sudan People Liberation Army.
SPLA-IO	Sudan People Liberation Army-In Opposition.
SPLM	Sudan People Liberation Movement.
SPLM-IG	Sudan People Liberation Movement - in Government.
SPLM-IO	Sudan People Liberation Movement - in Opposition.
TGONU	Transitional Government of National Unity.
UDSF	United Democratic Salvation Front.
UN	United Nations.
USSLM	United South Sudan Liberation Movement.
TROIKA	USA, Norway, and UK.
WUN	Western Upper Nile.

Author's Note

How President Kiir Betrayed Me
After President Museveni and I had saved him from being toppled by his former Vice President, Dr Riek Machar Teny-Dhurgon, President Kiir, ordered his Chief of Defence Staff, General Gabriel Jok Riak to arrest and fly me from Mayom County to Juba on May 31, 2018.

I was attacked in Mayom town on May 31, 2018, around 1200 hours by pro-government Mayom-based militias under the direct command of Major General Mathew Puljang Top. Three soldiers were killed, and two others wounded among my personal bodyguards who were accompanying me home. I did not fight in self-defense because it was not my intention to do anything stupid that would be used against me in a court of law. They arrested, shackled, and flew me to Juba the following day by a small caravan plane. On arrival at Juba International Airport, I was manhandled and thrown into a small national security van that did not have any ventilation, save for a few small holes created to let in some air. They threw me in an old underground bunker that was initially used as a torture chamber in Juba by the former Sudan military intelligence during the war to throw in and lynch suspected South Sudanese who were accused of being SPLA supporters. When Sudan Armed Forces withdrew to the North, the bunker became a dwelling ground for snakes, rats, and other harmful and poisonous creatures. The South Sudan People's Defence Forces (SSPDF) Military intelligence instead converted it for the same use against the South Sudanese accused of opposing President Kiir's regime. Many South Sudanese People, especially from Dinka Awiel, Nuer, and Equatorians have been lynched and tortured secretly in that bunker. Decomposed bodies could be smelt inside the bunker that is 4 by 8 meters long.

11

I was bundled into the bunker for three days with my legs and hands shackled and without food and water. I was forced to sleep on the rough floor and struggled to safety whenever I spotted a snake or a rat charging towards me in the dark, rolling on my belly from one side of the bunker to the other. I urinated in my pants for three days until when I was transferred from military intelligence custody under a cruel military intelligence officer, Brigadier General Kuel Garang Kuel to Military Police cells where the handcuffs were broken and removed because there were no keys to unlock them. The keys were thrown into a toilet by Brigadier Kuel Garang Kuel fearing that the handcuffs would be removed by his soldiers in his absence because the soldiers manning the custody were not happy with the way I was being humiliated despite being a comrade and the one who served and commanded various SPLA (SSPDF) units. "According to the South Sudan interim laws, "an accused person is presumed not guilty until proved in the court of law".

On June 16, 2018, the Chief of Defence Forces formed a committee under the late Maj. Gen. Michael Manok Kot to investigate my alleged rebellion against the government and whether I was in contact with the rebel leader General Paul Malong Awan. The charges were framed by Major General Mathew Puljang Top and some politicians from my home area, Bul Nuer, especially Tut Keaw Gatluak and Joseph Manytuil, the Governor of Unity State. Their major goal was to get rid of me politically and physically. The committee recommended to the former Chief of Defence Staff General Gabriel Jok Riak that they did not find me guilty of the alleged offences and suggested that I should be released immediately. However, the Chief of Defence Staff was very furious with the outcome of the committee, and he ordered the committee to do again the investigation until they find me guilty of the accusation. The committee redoing the investigation did a fresh investigation and this time round found me guilty and recommended that I should be arraigned in a General Court Martial (GCM) to answer the charges.

On January 3, 2019, the Chief of Defence Staff formed a General Court Martial that consisted of seven members namely:

1. Major Gen Thiik Achiek Hol   –   Court President
2. Brig. Gen Riak Tiop Deng   –   Member
3. Brig. Gen Duol Gony   –   Member
4. Brig. Gen Nhial Arou   –   Member
5. Brig. Gen Angui Geng   –   Member
6. Brig. Gen Hillery Oduho   –   Member
7. Brig. Gen Abubakar Mohamed –   Member

On February 4, 2019, I appeared before the General Court Martial and was asked if I had any objection to the formation of the members of the panel. I told the court that I had a reservation with the whole panel due to the following reasons:

Firstly, the SPLA Act 2009, Article 35 (4) says, in all cases, the presiding officer shall be senior in rank to the accused and that all members of the panel be at least equivalent in rank to the accused. Therefore, in my case, the presiding officer should have been a Lieutenant General and six Major Generals.

Secondly, the SPLA Act 2009, Article 36 (2), says that the Commander-in-Chief shall convene the General Court Martial when the person being tried is in the rank of a Brigadier General and above and the Chief of Defence Staff should convene a court in respect of other junior officers in the lower ranks. The court had been wrongly constituted and had to be dissolved in conformity with the above legal procedures. The General Court-martial was dissolved, and the Chief of Defence Staff recommended new names to the Commander-in-Chief to reconstitute a new General Court Marital, which the President reconstituted hereunder:

1. Major Gen Thiik Achiek Hol   –   Court President
2. Major Gen Atem Duot Atem   –   Member
3. Major Gen Peter Gatwech Gai –   Member

4. Major Gen Akuei Ajou Akuei –   Member
5. Major Gen Isiah Paul Lotole   –   Member
6. Brig. Gen Abubakar Mohamed –   Judge Advocate

The court conducted 17 sessions upon which it collected views and facts from the witnesses. The last session was held at a closed-door where closing arguments and mitigations were held. The court dropped some of the charges such as **Article 60** which related to treason and **Article 61** for offences relating to security that had been levelled against me earlier.

The only charge that remained was, disobeying lawful orders from your superior that were found in **Article 67** and Standing Orders **Article 69**, SPLA Act 2009. I was not given a fair trial because the General Court Martial was politically manipulated and received incentives in the form of bribery at the end of every session from the very people who masterminded my arrest. It is worth mentioning that, at the end of every session; a vehicle with the Presidential numberless plate from the house of the Presidential Security Advisor Tut Keaw Gatluak could be seen openly coming to the court to pick up some members of the GCM and dropping them at their respective homes. They could be seen giving a copy of that day's session to the driver of Tut Keaw Gatluak to take it to him so that he was informed about the day's court session. This had been continuing until when there was a serious tie vote among the members of the martial court because there was no substantial evidence found, that could be used to convict me. However, the tie was simply broken because of rampant poverty in South Sudan since the members of the GCM were bribed with hundreds of US dollars, according to one of the persons whom I will not mention here, who was among the team that went and paid the money to some powerful members among the members of the GCM who could influence the whole court.

14

The General Court Martial was forced to break a tie and improvised an arbitrary decision based on available administrative articles such as an **article 67** "disobeying of lawful orders from your superior" and article 69 "standing orders" SPLA Act 2009, which are very administrative and only can be applied in the case involving a junior SPLA personnel.

Therefore, the court martial endorsed the following ruling:

1. Demoted me from the rank of Major General to Private.
2. Dismissed me from the SSPDF to a civilian.
3. Sentenced me to one-year imprisonment from the date of my arrest on May 31, 2018, which elapsed during the trial.

After reading out the ruling, the General Court Martial asked me whether I would like to appeal against the verdict, which I turned down as there was nowhere, I could appeal against the verdict because the same powerful people who influenced and manipulated the decision of the court could also do the same in the outcome of my appeal. I gave up and left everything as it was decided by the General Court Martial and instead made the following statement: -

Honourable Court, when I moved from Wau to my home in Mayom, I asked for permission from the Commander of the Ground Forces on phone, who gave me five days and directed that I immediately proceed to Wau and hand over the command of 5th Infantry Division to the new Commander once I am back from my home. The Commander of the Ground Forces came to this Honourable Court and denied that he did not grant me permission to go home. I requested this Honourable Court to contact the Mobile Telecommunication Company (MTN) to provide this Honourable Court with my audio contact with the Commander of the Ground Forces, but this Honourable Court turned down my request in favour of the Commander of the Ground Forces.
I talked with the Commander - in - Chief, President Salva Kiir, in Mayom on the phone of General Akol Koor on the allegations that reached the President's office that I wanted to defect and join the rebels. But when the Commander-in-Chief found out that the allega-

tions were not true and that I only went home to offer traditional sacrifices on my father's grave, the Commander-in-Chief himself gave me five more days to spend at home before going back to Wau. The Commander-in-Chief told me that he would inform the Chief of Defence Staff that he had given me five more days to spend at home in Bieh-Nyang Payam.

I was attacked by Major General Mathew Puljang – I thought with orders from the SSPDF headquarters – and his soldiers killed three of my bodyguards and wounded two others. Interestingly, you have not arrested or demoted him to a private like what you have just done to me, yet I did not fight back because I knew that they are the government forces and I had to surrender to them knowing that they might have been ordered by the SSPDF headquarters to arrest me. It has appeared in this honourable court that nobody ordered Major General Mathew Puljang from the SSPDF to attack and arrest me. Major General Puljang was acting on his own without being ordered by the SSPDF headquarters to do so, but to my surprise, no action was taken against him by the SSPDF command. Instead, I am the one who has been arrested and charged with treason and other charges, while Major General Puljang was the one who killed three of my bodyguards and wounded two others, yet we did not fight them back in self-defence. Therefore, Mathew Puljang Top should be charged with murder. If I had fought Major General Mathew Puljang in self-defence, I wouldn't have been arrested as there would have been a lot of deaths on both sides in Mayom town, that is why my soldiers were only the ones killed and wounded in cold blood simply because I did not fight back.

Honourable Court, the verdict that you have just passed on me, that is a one-year imprisonment sentence starting from the date of my arrest on May 31, 2018, elapsed last May 2019, yet you are keeping me in jail. This is very humiliating to my personality.

Honourable Court, this case is politically motivated by a few politicians from my own community Bul Nuer who are conspiring and colluding with the Chief of Defence Staff to get rid of me politically

16

and physically. Therefore, I feel betrayed by these politicians and officers from my community in the face of the SSPDF command *despite the immense contribution that I have made in defence of this country and its constitution.*
End of My Statement!

The decision of the General Court Martial was supposed to be taken to the President for confirmation according to the Article 89 (2), SPLA Act 2009, that states, "The findings and sentence of a General Court Martial (GCM), shall be confirmed by the President and Commander -in-chief. In my case, the President did not confirm the decision of the General Court Martial, neither did he delegate a warrant to anyone to do so on his behalf or announced any Presidential decree on the South Sudan Broadcasting Corporation (SSBC) National TV as would be the case of a senior officer like I am.

It is worth mentioning that, the legal advisor in the office of the President the late Molana Majok Mading connived with Honorable Tut Keaw Gatluak and General Gabriel Jok Riak not to submit the decision of the General Court Martial to the President for confirmation, because of some legal irregularities involved in the court procedures which were likely to be in my favor if it were submitted to the President who would be much obliged by the law to dismiss the case altogether. They went ahead and announced the decision of the General Court martial at a military parade in Bilpam (SSPDF HQ), because the late Molana Majok Mading wrote to the Chief of Defence Staff in his capacity as legal advisor in the office of the President giving a greenlight to announce the judgement without any warrant signed by the President himself to confirm my conviction.

They went and briefed the President about what they had done illegally and by telling the people that the President was so busy with other national issues that he could not sign the confirmation letter by himself. The President kept quite without reacting to the illegality of the announcement, instead the President told them to wait and

17

see any negative reaction from the public or army so that he would pretend as if he was not aware of what was going on in the General Court Martial and he would eventually try to dismiss the case if things go wrong in the public or within the army.

Below is the letter written to the Chief of Defence Staff by the late Justice Majok Mading to give a greenlight to the Chief of Defence Staff to make the pronouncement of my conviction at a military parade ground as planned by the Chief of Defence Staff and Honorable Tut Keaw Gatluak. The letter reads as follows:

"Subject: Confirmation of the Verdict of General Court Martial (GCM) Verdict on Criminal Case No. 22/2018 of Maj. Gen. Stephen Buoy Rolnyang. I refer to the referenced letter Number GCM/J/351/2019, from the President of the General Court Martial Maj. Gen. Thiik Aciek Hol dated 21 October 2019, addressed to H.E. the President and Commander-in-Chief of the SSPDF, in which the General Court Martial (GCM) read out the verdict to the above-mentioned convict. According to the documents on record, and after the court ruling, the convict was asked as to whether he wishes to exercise his right of appeal or not in accordance with the provisions of Section 88 (2) of the SPLA Act, 2009, read together with Section 263 of the Code of Criminal Procedure Act, 2008. The convict is on record as having been satisfied by the verdict and hence did not intend to appeal.

Therefore, H.E. the President has confirmed the verdict of the General Court martial (GCM) in totality and he has further directed that the convict should be released forthwith".

Have the assurance of my highest regards and consideration.

Molana Majok Mading,
Legal advisor Office of the President.

After I had been released from the jail, I happily joined my family and stayed at home for three years and later decided to register and establish a security firm called **Jamus Security Service** (JSS) in Juba, at least to support my family financially so that I could no longer depend on the government support. When there was a contest for a position of the Governor of Unity State, I decided to submit my application to the office of the SPLM with an intention to be appointed to the position based on a popular demand from the people of Unity State to serve them as their governor to reconcile them, restore peace and rebuild the much-needed beautiful social fabrics, harmony and co-existence that used to be among the communities of Western Upper Nile that were destroyed by the SPLM infighting. Unfortunately, when some powerful members of the Sudan People Liberation Movement and National Congress Party (SPLM-NCP) found out that my application for the governorship position had reached the office of the SPLM and my name was about to be short-listed and subsequently be presented to the President for final endorsement, they went and talked to the former SPLM Secretary General Nunu Kumba and paid her handsomely and Kumba immediately removed my name in favor of Joseph Nguen Manytuil. They re-appointed Joseph Nguen Manytuil as the Governor for indefinite period while doing nothing for the people of Unity State. Instead, he is discriminating and killing the people of Unity State on clan lines. This is demonstrated in what has been going on in the South of Unity State areas of Koch, Leer and Mayendit, where thousands of people have been displaced and villages burnt down. The violence has also seen the killing of children and elderly people who could not run for their lives and all other forms of evil acts including gender -based violence.

The SPLM-NCP quest to get rid of me did not stop there. They had planned to assassinate me when Joseph Manytuil was finally re-appointed the Governor of Unity State. The population of Unity State reacted negatively and protested, but in vain against the re-appointment of Joseph Nguen Manytuil as the governor of unity State. The

SPLM-NCP concluded that it was me who instigated people of Unity State to protest the re-appointment of Joseph Nguen Manytuil. They then onwards, staged several assassinations attempts on me but failed because I maintained a low profile in Juba. It was while here that I received such an information about my possible assassination by the SPLM-NCP.

When I felt a high-level of insecurity targeted against me, I finally decided to sneak out of Juba to flee for my life and joined the movement of General Paul Malong Awan in Nairobi because I was pushed beyond the limit by the SPLM-NCP. When I found Malong's group was negotiating with the regime for potential peace agreement that would pave the way for his possible return to Juba to work again with the SPLM-NCP, I decided to pull out from Malong's group with the rest of my colleagues who defected with me from Juba and formed a new movement called the South Sudan People's Movement and South Sudan People's Army (SSPM/SSPA) with the aim of liberating the people of South Sudan from the SPLM-NCP politics of tribalism, hatred, injustice, repression, institutionalized massive corruption, bondage and to install a genuine democratic system of governance that would pave the way for the people of South Sudan to choose their own leaders in order to move the country forward.

In the SSPM/A, we have mobilized thousands of soldiers from SSPDF and other armed groups, youth, and intellectuals. Our forces are in Upper Nile, Unity State, Northern Bhar El Ghazal, Eastern Equatoria, and Raja, all along the border of South Sudan with Sudan and Uganda. We have set three phases to achieve our objectives one after the other.

Phase-1: Mobilisation and recruitment, which we have done successfully.
Phase-2: Organisation and training which is ongoing now.
Phase-3: To confront and challenge the regime physically and violently which was pre-empted by the regime when the Commission-

er of Mayom County in Unity State, Major General James Chuol Gatluak and Commander of 4th Infantry Division launched a coordinated attack on our location where we were conducting military training at a place called Bong on July 21, 2022, killing one soldier and wounded many others on our side. Our forces carried out a counterattack on Mayom town on July 22, 2022, in which the Commissioner of Mayom County died in the process when his grass-thatched house was hit by an RPG-7 and caught fire.

As a result, the National Security Advisor Tut Keaw Gatluak, brother to Late Commissioner James Chuol Gatluak paid more money to the Sudan Rapid Support Forces (SRF) to hunt down our senior officers in Sudan, something seen as a possible revenge for the death of his brother. On August 6, 2022, SRF members kidnapped four of our officers in the refugee camp in Fula state in Sudan and handed them over to the Governor of Unity State on August 7, 2022, through Heglig-South Sudan border. The Governor of Unity State Joseph Nguen Manytuil received our officers and summarily executed three of them by firing squad and put the fourth one, Late Gatluak Majiok Liey inside a grass thatched house and burnt him alive. The video of the executed officers went viral on the social media.

On August 15, 2022, the South Sudan National Security Advisor Tut Keaw Gatluak headed a high -level government delegation to Ethiopia that consisted of Director General of Internal and External Security and South Sudan Minister of Interior. Their visit related to the fact that, they thought I was present in Ethiopia, and they asked the Ethiopian authorities to hand me over to them as they did in Sudan and paid secret security agents to hunt me down either to liquidate or kidnap me in Addis Ababa or repatriate me to South Sudan.

Gen. Stephen Buay Rolnyang

Preface

The publication of this book is presented because of the different political arguments and views among the South Sudanese political and armed groups who argue that the root causes of the conflict in the Republic of South Sudan are tribal, and not political, but my argument is that the root causes of the conflict in the Republic of South Sudan are power struggle and leadership as explained in the introduction and the subsequent explanations in the book itself.

The first Power struggle among the South Sudanese liberators started in 1967, when Father Saturnino Lohure, who was the over-all commander of the Anyanya-I forces, was killed by the Ugandan forces while crossing the border to Uganda, and the remaining two Anyanya-I officers, Emilio Tefang, and Second Lieutenant Joseph Lagu, were at odds with each other, fighting over who could be the leader of the Anyanya-I command. Joseph Lagu broke away and formed his own movement called the South Sudan Liberation Movement (SSLM). He managed to regroup all the different camps of the Anyanya-I from other parts of the Southern regions, and in 1969, the Anyanya-I rebels managed to establish contact with Israel to supply them with arms through Ethiopia and Uganda, and to train the Anyanya- I recruits, and bought weapons from the Simba, a Congolese rebel group.

The second power struggle among the South Sudanese liberators started in 1983, when Colonel Dr John Garang De-Mabior and his colleagues Major Kerubino Kuanyin Bol, Major William Nyuon Bany, and Captain Salva Kiir Mayardit among others led the rebellion in Bor, Ayod and Malakal respectively and went to the bush and joined the Anyanya-II group that was already in the bush for three years. These were the ex- Anyanya-I soldiers who were completely disillusioned with the terms of the Addis Ababa peace agreement that was signed in 1972, between the government of Sudan and the leadership of the Anyanya-I liberation movement led by General Joseph Lago. The ex-Anyanya-I soldiers discontented with the peace agreement and decided to remain in the bush and called themselves the Anyanya-II, after the name Anyanya-I and established their base at Bilpam on Ethiopia-Sudan border.

The Anyanya-II group was led by former Anyanya-I politicians and officials namely Akuot Atem Mayen, Samuel Gai Tut, and Joseph Oduho among others. The Anyanya-II leaders wanted Akuot Atem to lead and retain their seniority over the Bor mutineers with the objective of fighting for the liberation of South Sudan while on the other hand the Bor mutineers wanted John Garang to lead and retain their seniority over the Anyanya-II group with the objective of fighting for the liberation of the whole Sudan, and as a result the fighting ensued between the two groups and the Bor mutineers prevailed and backed by the then Ethiopian government. The Anyanya-II group was dislodged from Bilpam in October 1983, and went to South Sudan territory where they decided to operate as independent group and sought military support from Sudan government to fight the SPLM/A, the movement which was later formed by the Bor mutineers. The root cause of the conflict between the Bor mutineers and the Anyanya-II was power struggle and ideological differences and not the tribal differences. Each group wanted to lead with different ideology. Bor mutineers wanted to fight for the unity of the whole Sudan while Anyanya-II group wanted to fight for the separation of South Sudan from the rest of Sudan. In fact, the Anyanya-II were called separatists and Bor group was called unionist. Most importantly, the leaders of two groups come from Dinka Bor, Akuot Atem of the Anyanya-II group and John Garang of Bor group.

The third power struggle among the South Sudanese liberators was in 1987, when Colonel Dr John Garang had a confrontation with his two senior members of the SPLM/A political high command namely Kerubino Kuanyin Bol and Arok Thon Arok. John Garang accused the two senior SPLM/A members of attempting to overthrow him and took action to arrest them. The two high command members accused John Garang of running the movement as his own private property and blaming him for failing to convene a meeting of the high command since the inception of the movement.

Note: *"There was no tribal conflict here as John Garang and the two high command members all hail from Dinka tribe. It was just power struggle".*

The fourth power struggle among the South Sudanese liberators started in 1991, when the second group of the SPLM/A members of the political and military high command: Riek Machar Teny Dhurgon, Lam Akol Ajawin and Gordon Koang Chuol announced a coup in Nasir on August 28, 1991. They accused John Garang of dictatorial tendency and running the movement like a briefcase business and prolonging the war of liberation struggle fighting for the whole Sudan. The Nasir coup was quickly taken advantage of by some unruly and rogue Nuer - Dinka militias and commanders on either side who misconstrued it as a fight between Dinka and Nuer. As a result, the conflict spilled over into a serious tribal confrontation between Nuer and Dinka whereby the two tribes took the law into their own hands by targeting members of each of the ethnic groups who were residing on either side of the tribal line and deadly cross border cattle raids continued for years.

Note: *As can be discerned from these developments, the Nasir coup was not tribal, it was power struggle within the SPLM/A. It is because of failed leadership in the two factions that allowed rogue and unruly commanders and militias to take the law into their own hands and target tribal members of either side on ethnic lines.*

24

The fifth power struggle within the top leadership of the SPLM ruling party started in 2013, when Salva Kiir Mayardit accused his Vice President Riek Machar Teny Dhurgon of attempting to oust him from power. The move spilled over into a military confrontation between the forces loyal to Salva Kiir Mayardit and Riek Machar and plunged the whole country into a civil war. As a result of total leadership failure, the unruly and rogue militias known as Mathiang-Anyor and some other uncivilized Dinka men in Juba took the law into their own hands by targeting Nuer in Juba. The conflict took tribal dimension and spread fast across the country causing serious destructions with civilians of either ethnic Dinka or Nuer being targeted on ethnic lines. It also caused displacement of more than two million people.

Note: *Lack of leadership qualities in the two leaders President Kiir and his Vice President Riek Machar contributed to the breeding of tribal nationalism and hatred between Dinka and Nuer ethnic groups. They failed to control their tribal militias and instead they armed them to fight for them and sustain their leadership struggle.*

Chapter I

Anyanya I: Power Struggle

Anyanya (also Anya-Nya) was a guerrilla name from a Madi word, which means snake poison (venom), which kills human beings slowly. It is considered one of the most significant guerilla movements that emerged in Southern Sudan in the 1950s.

According to the late Major Gen (rtd) Monani Alison Magaya, Anyanya-I was the first civil war in Sudan, which was sparked by an order from the Government of Sudan to transfer all Southern soldiers of the Equatoria Corps from South to North Sudan and be replaced by the northern Arab forces. The former Minister of Interior and head of South Sudan's embassy to Switzerland also explains that the war was fuelled by rumours that the southern forces would be disarmed by force before they could be moved to the north.

Making matters worse was the shooting of demonstrators of textile workers of Nzara in Yambio, which angered southern soldiers who were deployed in the area. The situation remained tense between the northern and southern forces, until August 18, 1955, when the mutiny broke out at Torit garrison, the headquarters of the Equatoria Corps, and spread to all other Southern regions.

Even though it was not properly coordinated, southern mutineers in Torit started killing the northern officers, administrators, merchants, and their families. They also killed some Southerners who were accused of being collaborators by cooperating with the Northerners against the Southerners. The remaining northern survivors ran and took refuge with the missionaries in the churches. The Southern mutineers and their families moved to the Uganda - Sudan border to settle their families in the refugee camps and would come back to fight the northern soldiers.

Attempts by the British and Sudanese governments to restore order in Torit and other parts of South Sudan failed. The mutineers who were arrested by the northerners were summarily executed in Torit-The Sudan army carried out revenge killings or retaliations on innocent civilians around the area, and this encouraged more Southerners to join the rebellion.

On January 1, 1956, the British government granted formal independence to Sudan after it persuaded Southern politicians to side with the northerners with the assurance that their quest for a federal constitution would be given serious consideration thereafter. Sudan became independent when so many issues had not been resolved at all. Parliament suspended the constitution in 1958 instead of allowing it to guide the decision on federalism. The referendum for the South Sudan that was supposed to take place was cancelled in 1982. Instead, the people of South Sudan registered their opposition to the subdivision of the Southern region.

There was disagreement over the drafting of the permanent constitution of Sudan by British experts on the question of whether the country would be a federal or a unitary state or should have a secular or an Islamic constitution. Southern politicians supported federalism to save South Sudan from being controlled by the powerful Muslim Arabs-dominated northern central government, while northerners feared the possibility of federalism leading to the separation of Southern Sudan from the rest of the country.

Southerners did not want to be "Islamized and Arabized" and the policy of this "Islamisation and Arabisation" was the main cause of armed struggle in South Sudan during Ibrahim Aboud's tenure. Many Southerners were encouraged to be converted to Islam, especially the students.

In 1964, Christian missionaries were blocked and expelled from South Sudan, and the military campaign was stepped up to hunt down the Anyanya soldiers who kept hiding in the bushes because they did not have the capacity and lacked support to sustain the fight. The army resorted to burning villages, accusing the villagers of harbouring rebels. Such repressive campaigns targeted Southern officials, and this encouraged many educated Southerners to remain opposed to the government.

Many civilians were arrested and tortured, and this only succeeded in sparking more defections of senior southern political figures to join the rebellion namely, Fr. Saturnino Lohure, Aggrey Jadden, Joseph Oduho, and William Deng Nhial. They went and formed the core of a guerrilla movement in exile and called it the Sudan African Nationalist Union (SANU). The Anyanya fighters lacked external military support.

They depended mainly on what they captured from the enemy after laying ambushes to their patrols across the Equatoria region. In October 1964, Gen. Ibrahim Abboud stepped down after a series of demonstrations in Khartoum. He was replaced by a caretaker civilian government. Anyanya leaders were invited to attend a round table conference offered by the new civilian rule to legalize the formation of political parties.

Those of Clement Mboro formed a political party called the Southern Front (SF) in Khartoum under Clement Mboro. Clement himself was appointed Minister of Interior in the new caretaker government. When the new civilian caretaker government invited the Southern politicians to the round-table conference, there was a split within SANU. The invitation centred on discussions around the problem of South Sudan.

William Deng Nhial agreed to attend the conference and became the leader of SANU inside Sudan, advocating for a federal solution. Those of Aggrey Jadden and Joseph Oduho returned to Uganda as leaders of SANU, outside Sudan, advocating for the separation of South Sudan from the rest of Sudan.

28

In 1967, Fr. Saturnino Lohure who was commanding Anyanya forces that mutinied from Torit, was killed by Ugandan forces while crossing the border to Uganda. After the death of Fr. Saturnino, two army officers namely, Emilio Tefang, and second lieutenant Joseph Lagu, started fighting over the leadership of Anyanya-I.

As a result, second lieutenant Joseph Lagu broke away and formed his own movement called the South Sudan Liberation Movement (SSLM) after which he was joined by Dinka leaders such as Gordon Muortat Mayen and Akuot Atem.

Joseph Lagu managed to regroup all the different camps of the Anyanya-I from other parts of the Southern regions, and in 1969, the Anyanya-I rebels established contact with Israel to supply them with arms through Ethiopia and Uganda and to train the Anyanya-I recruits. In addition, the Anyanya leadership managed to buy weapons from the Simba, a Congolese rebel group. On May 25, 1969, Jaafar Nimeiri took power in Khartoum and declared that there was no military solution to the South Sudan political crisis, and on February 27, 1972, South Sudan Liberation Movement negotiated peace with the Sudan government under Nimeiri in Addis Ababa, Ethiopia.

The South Sudan Liberation Movement (SSLM) accepted the proposed regional autonomy that was proposed and agreed to the integration of the guerrilla forces of the Anyanya-I into the national army and other organized forces. The agreement granted the Southern regional government powers to raise revenues from local taxation to be added to the revenues from the central government.

In the negotiations also, SSLM proposed that the Southern soldiers be deployed in Southern Sudan for the protection of civilians from the Northern army. The Addis Ababa agreement was incorporated in the permanent constitution of 1973, and in May 1983, Nimeiri subdivided South Sudan into three regions of Bhar El Ghazal, Equatoria, and Upper Nile regions. As a result of the Anyanya-I war, two million people died of war, famine, and diseases and about four million people were displaced to refugee camps.

Anyanya II

The name Anyanya -II was created by the holdout groups or ex–Anyanya -I soldiers who expressed discontent with the terms of the Addis Ababa agreement. These ex-Anyanya -I veterans remained in the bush in their own camps on the Ethiopian–Sudan border, but they were not active until they were joined by various mutineers in 1976, led by second Lieutenant Vincent Kuany Latjor and Corporal Bol Kur from the Sudanese army garrison of Akobo.

The Ethiopian authorities used these Anyanya- II soldiers on their border to pressurize the Sudan government into stopping the supply of weapons to Eritrean rebels who were fighting the Ethiopian government from their base in Sudan.

The Ethiopian government promised to give full support to the Anyanya -II remnants on the Ethiopian border who refused to accept the Addis Ababa peace agreement. Nimeiri did not stop supplying the Eritrean rebels with ammunition and weapons as requested by the Ethiopian government and as a result, the Ethiopians continued to support the Anyanya -II by arming and training them. At the same time, Libya also started supplying the Anyanya-II group with arms and other military support through Ethiopia. Anyanya-II remained insignificant in Ethiopia until 1980 and 1981 when they gained more support from Southern soldiers and civilians.

In 1982, the Anyanya-II group managed to establish bases in Nasir, Bentiu, and Fangak districts and established contact with the local Southern police and soldiers persuading them to join the rebellion. By that time, Akuot Atem De Mayen and other Southern politicians who refused to accept the Addis Ababa agreement were in Ethiopia seeking political and military support to fight the government of

Sudan and liberate South Sudan.

Akuot Atem from Dinka Bor was appointed as Minister of Interior in the Anyanya-I government. Samuel Gai Tut from Lou Nuer was appointed Minister in the interim Southern Sudan regional government with William Abdalla Chuol from Fangak Nuer.

Samuel Gai Tut and William Abdalla Chuol were still working with the government and were in contact with Akuot Atem who was the leader of the Anyanya-II on the Ethiopia - Sudan border. Throughout this period, Samuel Gai Tut and William Abdalla Chuol were smuggling arms and ammunition to the forces of Akuot Atem.

Samuel Gai Tut was caught and charged with supplying the rebels with arms and ammunition. (Douglas Johnson, 2003). Samuel Gai Tut and William Abdalla Chuol defected and joined Akuot Atem on the Ethiopian border and formed their movement called the United South Sudan Liberation Movement (USSLM), with the objective of pushing for the separation of South Sudan from the rest of Sudan. Akuot Atem Mayen was nominated as the leader of the USSLM. By that time, there were two Anyanya -II camps on Ethiopian border, the political wing of USSLM based in Iteng headed by Akuot Atem De Mayen, Joseph Oduho, Samuel Gai Tut, William Abdalla Chuol Deng, while the military wing was based in Bilpam under the command of Gordon Koang Chuol, the overall commander of the Anyanya -II forces at Bilpam.

Later, they were joined by second lieutenant Vincent Kuany Latjor and his group who defected from Akobo Sudan Army Garrison in 1976. The Anyanya-II forces at Bilpam were redundant, they were not active or operational.

They just used to come down from the Ethiopian border to Sudan to forcefully snatch cattle and other valuable things from innocent civilians and take them to Bilpam where they used the cattle and

31

other materials for paying dowries.

As a result of these despicable acts, the group accumulated a lot of wealth and built big grass-thatched houses and made Bilpam their permanent home and forgot about the liberation struggle that took them to Bilpam. They did not challenge any single Sudan Army Garrison until they were joined by the Bor mutineers in 1983.

Note: *The main cause of the problem in the Anyanya-I leadership between Joseph Lagu and Emilo Tefang was a power struggle after the death of Fr. Saturnino and was not a tribal since the two guys were all from the Equatoria region.*

Anyanya-I Veterans

Figure 1. Joseph Lagu left and David URI, an Israeli soldier in the Middle 1960-1970

Figure 2: Anyanya-I expedition between Juba and Torit. 1960-1970.

Chapter 2

Anyanya-II and Bor Mutineers' Ideological Divide

The mutiny by the Sudanese Armed Forces (SAF) battalion 105 of the 1st Infantry Division in Bor in Jonglei region on May 16, 1983, where a fight ensued at Malual Chaat army garrison between the mutineers (rebels) and loyal troops marked the beginning of the second Sudanese civil war. It was largely a continuation of the first Sudanese civil war of 1955 to 1972. During this battle, about 78 soldiers were killed from both sides, including one Major and seven Non-Commissioned Officers (NCOs), and other ranked soldiers.

The mutiny that was sparked by an order directing the transfer of Battalion 105 from Bor in South Sudan to Sudan (North) was led by Major Kerubino Kuanyin Bol. Jaafar al-Nimeiri, then president of Sudan sent Colonel John Garang De-Mabior as a Dinka to convince his fellow tribesmen in Bor town to lay down their arms.

However, unknown to Nimeiri, Garang was already in secret talks with Major Kerubino Kuanyin Bol. Garang arrived in Bor on May 14, 1983, and joined the mutineers. Once there, he instructed the mutineers to collect and carry weapons, food, and medicine and moved to the jungle towards the border of Sudan with Ethiopia. They moved to the border of Ethiopia and set up their base there under his command. They established contact with the soldiers in Akobo, Pochalla, Pibor, Kapoeta, Rumbek, and Aweil garrisons, urging them to join the rebellion.

After a few months, the soldiers, other organized forces, and government officials from these regions defected and joined the rebellion. It is worth noting that the bulk of the forces that formed the first Sudan People's Liberation Army (SPLA) battalions were from Nuer, an ethnic group mainly concentrated in the Greater Upper

34

Nile region and who also live in the Ethiopian region of Gambela. The combination of the attack on Bor and the subsequent abolition of the South Sudan region created further rebellions and desertions from various garrisons across South Sudan. In July 1983, about 3,000 soldiers defected and joined the new guerrilla base, which had been established on the Ethiopia border.

A further 1,000 soldiers defected from Bahr El Ghazal's various army garrisons and established their bases in the countryside in the rural villages and their leaders went to meet with Colonel Garang on the Ethiopia border. When Garang and his group arrived on the Ethiopian border, they were received and welcomed by the ill-equipped Anyanya-II forces that were already in Bilpam and Iteng areas. The fact that Anyanya- II soldiers had already fought these Bor mutineers before they joined the rebellion, made it difficult to unite them.

The Anyanya-II officers claimed that they were senior to the officers who had just joined them on the Ethiopia border from Bor. In July 1983, Colonel Garang held a meeting with United Southern Sudan Liberation Movement (USSLM) leaders and Anyanya-II commanders on the possibility of unifying the two movements to fight one common enemy – the Sudan Government.

All the commanders agreed in principle to unify the movement, but they failed to agree on the name of the new movement and its leadership structure. Anyanya -II commanders proposed South Sudan Liberation Movement/ Army (SSLM/A) to be the name of one unified movement to help liberate South Sudan from Arab rule, while Colonel Garang's group proposed Sudan People's Liberation Movement/Army (SPLM/A).

When the Ethiopian authorities were informed that there was a disagreement between the two factions, the leaders of the USSLM/A and SPLM/A were invited to Addis Ababa to meet Ethiopian senior officials. Colonel Garang, Akuot Atem, Joseph Oduho, Samuel Gai

Tut, and Salva Kiir Mayardit were airlifted from the Mangok area to Addis Ababa by helicopter courtesy of the Ethiopian Government. When they arrived at the meeting, they were asked by the Ethiopian authorities to present their respective movement's manifestos that explained the objectives and goals of their different organizations.

Colonel Garang's faction had recommended that the movement should keep on fighting for the unity of Sudan. Garang was aware of the regional politics and that the Ethiopian Government was supporting a united Sudan and intended to overthrow President Nimeiri due to his open support to the Eritrean rebels who were fighting to separate Eritrea from Ethiopia.

If Colonel Garang had adopted the ideology of the Anyanya-II that called for the separation of South Sudan from the rest of Sudan, the Ethiopian Government officials at the time would not have supported them since it would look like Ethiopians were endorsing the push by Eritrean rebels to separate from the country.

Colonel Garang and his rebel faction from Bor were at odds with the old Anyanya-II faction of Akuot Atem, Samuel Gai Tut, William Abdalla Chuol Deng, and Joseph Oduho. This team of Anyanya-II veterans did not buy Garang's overall strategy. They were keen on fighting for the liberation and independence of South Sudan.

The Anyanya-II faction wanted to retain their seniority over the Bor mutineers, and they preferred the separation of the military from the political wing of the movement as they did in the Anyanya-I, where the political wing was separated from the military wing.

Samuel Gai Tut and the other members of USSLM proposed that Akuot Atem be the leader of the newly created political wing – the Sudan People's Liberation Movement (SPLM) and that he would be deputized by Joseph Oduho. They further proposed that Samuel Gai Tut be the leader of the military wing – the Sudan People's Liberation Army (SPLA) and that Colonel Garang become his deputy.

These suggestions were rejected by Kerubino Kuanyin Bol and William Nyuon Bany who viewed Samuel Gai Tut and his faction as their enemies before they joined the rebellion. The two wanted Garang to be their leader, while old Anyanya-II veterans who also claimed that they were senior to Garang insisted it was wrong for him to be named as their leader.

After listening to their presentations and scrutiny, the Ethiopian authorities chose the one that was presented by SPLM/A and asked the Akuot Atem's group to abandon their movement's ideology and join Colonel Garang's faction. The Ethiopian officials admired Garang because of his education and his youthfulness compared to Anyanya-II veterans who had little education and were older.

Akuot Atem, Samuel Gai Tut and Anyanya-II senior officials agreed in principle and adopted the new name (SPLM/A), but raised concern on the leadership, ideology, and policy of the movement. Samuel Gai Tut and Akuot Atem went to their camp at Iteng away from Colonel Garang's camp. All this time Garang's faction was secretly getting weapons and ammunitions as well as receiving trainings from the Ethiopian authorities. The USSLM camp in Iteng noticed this and withdrew to South Sudan with their supporters on September 3, 1983, and operated independently, for fear of being attacked and arrested by the Ethiopian forces. The USSLM forces established bases in Waat and Kongor areas, the respective home areas of Samuel Gai Tut and Akuot Atem. They instructed Anyanya-II commanders at Bilpam not to recognize SPLM/A under Colonel Garang and his group.

In October 1983, the Ethiopian Government and Bor mutineers, mainly from SPLM/A launched an attack on the Anyanya-II camp at Bilpam and dislodged, dispersed, and disorganized them. The SPLA scattered the Anyanya-II forces and incorporated some of them into its structure. Commander Duay Taitai Badeng from Bul Nuer was killed in the attack on the Anyanya-II side. Vincent Kuany Latjor and his supporters surrendered to the SPLM/A. Gordon Koang Chuol withdrew with the bulk of the Anyanya-II forces and joined those of Akuot Atem, Samuel Gai Tut and William Abdalla Chuol group in South Sudan territory and the two groups unified their command and called it Anyanya -II under the command and leadership of Akuot Atem De Mayen. After some time, a serious split emerged again within the newly unified Anyanya-II faction. Unfortunately, both Samuel Gai Tut and Akuot Atem De Mayen were mysteriously killed by the end of 1983. Commander William Abdalla Chuol took over the leadership of the Anyanya-II and Gordon Koang Chuol was named his deputy. The Anyanya-II new leadership that was mainly made up of Nuer sons established contact with the Sudan Government and they started receiving military supplies from Khartoum.

The Sudan Government created several militias as friendly forces against the SPLA, especially amongst the Toposa, Murle, and Mundari communities. It also intended to nurture a Nuer army to fight the SPLA, which they saw as a faction that was dominated mainly by the Dinka people. Anyanya-II drew its support from the Bul Nuer people in Western Upper Nile under the command of General Paulino Matip Nhial, Laak Nuer in Fangak under William Abdallah Chuol Deng, and from Lou and Jikany Nuer. Anyanya -II forces occasionally attacked and dispersed SPLA recruits and refugees who crossed from Bahr El Ghazal, Western Upper Nile, and Fangak under the SPLA escort to Ethiopia's border. They cut the SPLA supply line from Ethiopia to Fangak, Western Upper Nile, and Bahr El Ghazal areas. The fighting between Anyanya -II, and SPLM/A continued for five good years until Gordon Koang Chuol joined the SPLM/A in 1988, along with the bulk of the Anyanya-II forces.

> Note: *It is worth noting that the differences between the Anyan-ya-II veterans and the SPLM/A were based on ideology and personal differences, and were not tribal, since Colonel John Garang and Akuot Atem were both Dinka from Twic Bor clan in Kongor District and commanders from both Nuer and Dinka were supporting either side to sustain themselves in power.*

Anyanya-II Veterans

Figure 3: Uncle Joseph Oduho

Figure 4: Samuel Gai Tut

Figure 5: Vincent Kuany Latjor

Figure 6: Gordon Koang Chuol

Chapter 3

SPLM/A Power Struggle

The Sudan People Liberation Movement and Army (SPLM/A) was formed in 1983, with the ideological objectives for the liberation of the whole of Sudan to create a New Sudan, a "united, democratic and secular Sudan" where equitable management of diversity and respect for identities and cultures of all "national" groups would prevail.

However, things did not go well as planned. With time, members of the SPLM/A political - military high command began to lose hope about the realization of the ideology of fighting for the liberation of the whole of Sudan. There was therefore a need for them to review the manifesto of SPLM/A to change the ideology from fighting for the whole Sudan which they thought would take time and resources. They resolved to change the plan and ideology and fight only for the liberation of Southern Sudan including Abyei, Nuba Mountain, and Southern Blue Nile.

The first power struggle within the SPLM/A started in 1987, when Colonel Dr John Garang had a confrontation with his two senior members of the SPLM/A Political Military High Command (PMHC): Kerubino Kuanyin Bol and Arok Thon Arok. John Garang accused the two of attempting to overthrow him and directed that they be arrested. The two high command members accused John Garang of running the movement as his own private property and blamed him for failing to convene a meeting of the high command since the inception of the movement. Second power struggle within the SPLM/A occurred in 1991, when the members of the SPLM/A Political- Military High (PMHC) asked John Garang to convene an urgent high command meeting in June 1991, following the collapse of Mangisto regime and the changing regional and international situation to review and analyze the impact of the changes and to adjust

41

manifesto of the SPLM/A.

However, John Garang who had known about the whole plan avoided convening the meeting and instead ordered for mass mobilization campaign of the SPLA forces to converge around Juba to attack and capture Juba. The three high command members in Nasir, Commander Riek Machar Teny Dhurgon, Commander Lam Akol Ajawin and Commander Gordon Koang Chuol refused to contribute forces to go to Juba. They instead decided to act without waiting for the rest of the members of the SPLM/A Political - Military High Command who were sharing the same ideas and senior to them. On August 28, 1991, they sent a radio message to all the SPLA units that they had taken over the leadership of the SPLM/A and accused John Garang of dictatorship including running the movement alone. They had pledged to promote and advocate for human rights within the movement and called for the self-determination (**New ideology**).

The ouster of John Garang was quickly supported by Nuer militias that were fighting alongside the SAF against the SPLM/A, especially the Bul Nuer and Lou Nuer militia units operating in Mayom and Doleib Hill. The two SPLA factions were referred to as SPLA Nasir and SPLA Torit or SPLA mainstream. No side was willing to cede ground, leading to regular confrontation of forces in the main southern towns. Some of the deadly clashes took place in areas around Waat, Ayod, Sobat, Bor, and Bahr El Ghazal. There was also reported cases of a split among the Shilluk soldiers, many of whom refused to join Lam Akol while Maban soldiers that were under the command of Lam Akol also refused to join him and withdrew to the areas controlled by John Garang's supporters. Dr Garang ordered William Nyuon to attack Ayod, Adok, and Leer, but the Nasir faction forces in Ayod were reinforced by the Anyanya -II forces from Fangak and repulsed an SPLA attack on Ayod commanded by William Nyuon Bany. Riek Machar sent a combined force of Anyanya -II and Nasir faction supported by armed Nuer civilians known as "White Army" to attack Kongor and Bor, the home area of John Garang.

The attack resulted in the massacre of innocent people in Bor and when Riek Machar was informed of the killings, he ordered immediate withdrawal of his forces from Bor to Ayod, Duk, and Yuai areas. As the fighting was continuing, SPLA Nasir faction was receiving military supplies from the Sudan government.

In January 1992, Paulino Matip Nhial, the Anyanya-II leader and his forces mainly made up of Bul, Lou, and Fangak, declared their allegiance to the Nasir faction. The Anyanya-II forces in Western, Eastern, and Central Upper Nile were all integrated into the SPLA Nasir faction. Paulino Matip was integrated into Nasir faction as full commander and appointed as a member of faction's political military high command as well as overall commander of Western Upper Nile zonal command. The Sudan government, with support from the SPLA Nasir faction under the command of Riek Machar recaptured more territories in Jonglei and Eastern Equatoria regions. Riek Machar released political prisoners, including Joseph Oduho, Kerubino Kuanyin Bol, and Arok Thon Arok, and in September 1992, William Nyuon Bany defected from the SPLA mainstream, switching his allegiance to the Nasir faction.

On November 14, 1992, the Nasir faction attacked Malakal town to capture it so that Riek Machar could gain support from regional and international community and win over forces of Torit faction to join him by justifying his plans of making democratic reforms in the movement and to be seen to be fighting for the liberation of South Sudan, contrary to Dr Garang's vision.

In 1993, Riek Machar, Kerubino Kuanyin Bol, William Nyuon Bany, Joseph Oduho, Arok Thon Arok, and other senior members of Nasir faction high command met and re-named the Nasir faction as the SPLA-United which they later changed to South Sudan independent Movement/Army (SSIM/A). They proceeded to Panyagoor, the home area of Arok Thon Arok as part of their mobilization, but they were attacked by the SPLA forces of Torit faction at Panyagoor where Joseph Oduho and Commander Kuac Kang were killed in the attack and Riek Machar was defeated and withdrew to Waat area.

On March 13, 1994, SPLA Nasir faction forces in Lafon captured a government convoy that was passing through their area and afterwards they persuaded William Nyuon to rejoin the SPLM/A, and on April 27, same year, Dr Garang and William Nyuon signed what they called the Lafon Declaration that paved the way for Nyuon's rejoining the SPLM/A Torit faction which marked the reunification of the movement.

When Riek Machar heard about the declaration, he rejected it and his reaction annoyed Nyuon who later announced that he had dismissed Riek Machar from the SPLA Nasir faction and that the movement had been reunified with the SPLM/A Torit faction. Nyuon formed a new Executive Council, and he was supported by the SPLM/A Torit faction in a joint successful offensive against the government and the SPLM/A Nasir areas where William Nyuon fell into an ambush and was killed in January 1996.

Collapse of Khartoum Peace Agreement

When Riek Machar failed to secure military support from the international community to fight the government of Sudan and SPLM/A mainstream, he opted to sign peace deal with the Khartoum government on April 21,1997 and was appointed President of Southern Sudan States Co-ordinating Council. He was also made Commander-in-Chief of the South Sudan Defense Force (SSDF) by other South Sudanese armed groups that were fighting alongside the Sudanese government against SPLM/A under the umbrella of the political wing of the United Democratic Salvation Front (UDSF). The agreement defined a four-year interim period to enable the Southern states recover from the civil war and formed a coordinating council of the Southern states to oversee the transition government.

Under the agreement, a referendum was to be held after four years to enable South Sudanese decide to remain in a unified Sudan government or separate and form their own South Sudan government in accordance with the boundaries that were drawn in 1956.

The agreement, however, bestowed the control of the armed forces and other security apparatus in the hands of the central government, while the government of the Southern states was allowed minimal control over economic development only. The latter was, clearly, non-existent. The President of the coordination council was appointed by the President of the Republic, while the president of the southern coordinating council appointed the cabinet ministers and governors of the Southern states through final approval by the President of the Republic. Unfortunately, this agreement was not implemented in letter and spirit as had been agreed earlier. Riek eventually got frustrated and resigned and went back to the bush.

Gen Paulino Matip Falls Out With Riek Machar

In September 1997, a misunderstanding erupted between the forces loyal to Riek Machar and those loyal to General Paulino Matip over the appointment of the Governor of Unity State. Dr Machar wanted Taban Deng Gai to be the Governor, while General Paulino Matip was in favor of favours Paul Liyliy Mathoat to be the Governor of Unity State. Following this disagreement that led to a fight, General Paulino Matip defected from the South Sudan Defence forces (SSDF) and formed his own movement and named it South Sudan United Movement and South Sudan United Army (SSUM/SSUA). The Sudan government under Omer Hassan El Bashir continued to supply both General Paulino Matip and Riek Machar with ammunition and guns to fight each other's forces in the battlefield which, generally, were rapidly dwindling in numbers and commitment.

On September 5, 1999, Commander Peter Gatdet Yaka, the overall operations commander of Paulino Matip forces defected from the South Sudan United Movement and Army (SSUM/A) led by Paulino Matip Nhial and joined the SPLA mainstream with the bulk of

Paulino Matip forces, leaving Paulino Matip only with a few body-guards inside his own compound. Peter Gatdet took control of Bul and other areas in Western Upper Nile and announced himself as commander of Western Upper Nile reporting to the Headquarters of the SPLA mainstream. Paulino Matip was overthrown and replaced by Peter Gatdet as a new militia leader in Bul Nuer and other near-by Nuer sections of Jagei, Leek and Jikany who also switched their allegiance to Peter Gatdet. Peter Gatdet continued to attack Dok, Haak and Nyuong who were still supporting Riek Machar. Fighting enraged again between forces loyal to Peter Gatdet and Riek Machar with destruction of lives and properties and massive displacement in southern Unity State with girls and women abducted, cattle and other properties of economic values were looted.

Kerubino Kuanyin Bol Forms Own Militia

Kerubino Kuanyin Bol was made Deputy Chairman and Com-mander-in-chief of the SPLA-United when he joined Riek Machar on April 5, 1993, a name which later changed to South Sudan Independent Movement/ Army (SSIM/A). He moved from Eastern Upper Nile to Western Upper Nile where he proceeded to Bahr El Ghazal with his own forces that were mainly made up of sons and daughters from Dinka Bahr El Ghazal.

However, he was defeated by SPLM/A forces in Northern Bahr El Ghazal, forcing him to withdraw to Abyei government army garri-son, where he established his headquarters and continued to oper-ate and recruit soldiers from Bhar El Ghazal mainly from Warrap. He was, however, reinforced by the forces of Paulino Matip from Bul Nuer militias.

But in 1995, Riek Machar dismissed Kerubino Kuanyin Bol, William Nyuon Bany, and Arok Thon Arok from South Sudan Independent Movement/Army (SSIM/A), accusing them of secretly signing mili-tary and political agreements with Sudan government to overthrow him and to be replaced by Kerubino Kuanyin Bol.

46

In January 1998, Kerubino Kuanyin Bol's forces attacked and seized Wau town, the capital of Bhar El Ghazal region briefly before his forces were driven out by Sudan Armed Forces. Kerubino Kuanyin Bol then announced he was rejoining the SPLM/A mainstream led by John Garang. Although Garang welcomed him back to the movement he was just assigned to a headquarters position rather than a field assignment, a position which he felt was humiliating and not representative of his rank or position in the society.

Kerubino Kuanyin Bol then decided to defect again, and he fled from Nairobi where a gunfight between two factions of the SPLA had earlier exposed the sharp divisions in the main opposition fighting to oust the Khartoum government. General Kerubino Kuanyin Bol fled to Mayom in Western Upper Nile where he stayed with his son-in-law, General Paulino Matip Nhial who married his daughter. He was later killed in a mysterious circumstance in Mankien by forces of Peter Gatdet Yaka on September 10, 1999, when Peter Gatdet defected from Paulino Matip camp to the SPLA mainstream.

Note: *The death of Kerubino Kuanyin Bol was coordinated and handled by the top brass of the SPLM/A leadership who instructed Peter Gatdet to eliminate Kerubino Kuanyin Bol.*

Riek Machar Resigns and Rejoins SPLM/A

In 1999, Riek Machar left Khartoum for Nairobi to regroup his followers who were scattered and some of them had already rejoined the SPLM/A mainstream, Torit faction leaving him with fewer followers and without a proper movement. He resigned in 2000, from the Sudan government as President of Southern Coordinating Council, formed his own faction under the name Sudan Peoples' Democratic Front (SPDF), and established his base at Maiwut near the Ethiopia border. In 2002, Riek Machar rejoined forces with the SPLM/A and became number three in the SPLM/A. He was below

President Salva Kiir Mayardit in the hierarchy of the SPLM/A.

Commander Gatdet Defects from SPLA

In 2002, Senior officers under Peter Gatdet, Commander John Jok Nhial and Commander Samuel Gai Yirchak, supported by civil servants who were working in Mayom County, conspired against Peter Gatdet and wrote a letter to Dr John Garang that they did not want Peter Gatdet to be their commander. They further demanded that Peter Gatdet be replaced forthwith with Commander John Jok Nhial and Commander Samuel Gai be appointed as his Deputy. The information reached John Garang's headquarters, which supported the idea, but he did not discuss the matter with Peter Gatdet. Instead, he called Peter Gatdet to his headquarters in Newsite and told Peter Gatdet that he should attend the school of adult education for his own good and that Peter Gatdet should recommend one of his senior officers to act in command until such time when he should go back to his command. Peter Gatdet refused to attend the adult education as advised by Dr John Garang' and insisted that he should continue working at his base.

John Garang played down the issue until later time when it would be addressed. Commanders John Jok Nhial and Samuel Gai Yirchak convened a meeting at Riah Village at 6pm in the evening and they invited me as an observer since I was based in Bahr El Ghazal and just came to Western Upper Nile on permission. During the meeting, Commander John Jok Nhial and Commander Samuel Gai Yirchak told us that they had taken over the command of Western Upper Nile from Peter Gatdet and threatened to deal with him if he sets foot in Western Upper Nile.

The pronouncement was followed by a moment of silence in the room as some of us present looked down in surprise. At this juncture, I raised my hand and told the two commanders that what they were doing was not in order in line with the military procedures. "It would be better for them to raise the complaint with the

SPLM/A leadership and wait for further directive from the Commander-in-Chief. Instead, they wished me away, saying I should stop interfering with their internal affairs. They added that I had just come from Bahr El Ghazal and therefore had no right to talk about their command affairs. The two Commanders further accused me of indirectly supporting Peter Gatdet.

I decided to keep quiet until the meeting ended after which I went back to where I was living. The same night, the information regarding the meeting reached Peter Gatdet's supporters, including Commander Tito Biel Wic who was at the frontline at Wangbieth area. Tito Biel Wic was a commander based at Peter Gatdet's headquarters, but he was not informed about this meeting, and this caused tension within the camp at night when he came to hear of it. The following morning, Commander Tito Biel Wic came to Riah with a sizeable battalion, and on seeing him, all the commanders in Riah ran helter-skelter fearing for their lives.

As others took off, I remained behind, met, and asked Tito Biel Wic, why he had come with such a big number of forces to Riah all the way from his base. Instead, Tito Biel ordered his forces to arrest me. I got angry and I ordered my bodyguards to get ready to fight Tito Biel Wic in self-defense, but the officers who came with Tito Biel, including William Bajuoy Makuet, intervened and stopped the commotion. I then ordered Tito Biel Wic and his officers to go and sit under a tree for a meeting to listen to their grievance since to my knowledge I had no problem with him and his unit. Tito Biel and his officers agreed and went under a tree where I joined and explained to them exactly what transpired during the previous night's meeting and blamed them for overreacting based on false information.

Tito Biel Wic told me that he had come to arrest the commanders who wanted to overthrow Peter Gatdet plus "you General Stephen Buay because you were also in the same meeting," he said. He said that First Lieutenant William Bajuoy Makuet, a relative of Peter

Gatdet, had sent him a message the previous night and informed him wrongly that all Commanders in Riah had been asked to report to SPLA headquarters, but they refused to go and instead wanted to overthrow Peter Gatdet. It was at this point that Commander Tito Biel Wic was asked by William Bajuoy Makuet to come to Riah to arrest the commanders who had convened the meeting if indeed he was not part of the group who wanted to oust Peter Gatdet.

I told Tito Biel Wic and his officers that, in fact, I attended the meeting, but I was not part of the conspiracy against Peter Gatdet. I told them further that I was in the meeting to help arbitrate based on the issues that were raised. By this time, as we were meeting under the tree, all the commanders who ran away were rounded-up in the bush by the forces of Tito Biel Wic and were brought back handcuffed. I ordered that they should be set free and taken to their respective homes to await further action on the matter the following morning.

I immediately established contact with General Bior Ajang Duot to send a chartered plane to evacuate the commanders to the headquarters at New Site before information could reach Peter Gatdet, otherwise, all of us would be killed, as he would not believe even my neutrality in this matter. I proceeded and blocked all radio communication lines that were directed to Peter Gatdet to ensure that no one talked to him regarding this matter. The next day, General Bior Ajang sent a plane early in the morning and I boarded the plane together with John Jok Nhial, Samuel Gai Yirchak, John Puoljor Wicyoak, James Nhial Wathkah, Karlo Kuol Ruac, and Philip Bipean Machar heading to New Site headquarters where the Commander-in-Chief of the SPLA John Garang' De Mabior was based. When we arrived at New Site, Peter Gatdet was called by the Commander-in-Chief to come and help resolve the situation between him and his commanders. He, however, refused to come, since he was misinformed that he would be arrested if he came. The Commander-in-Chief told Peter Gatdet, that he wanted Commander

Stephen Buay to go back to Western Upper Nile to act in his absence as a neutral person until the case between him and his commanders was resolved.

Peter Gatdet became furious and refused to accept the proposal made by the Commander-in-Chief, arguing that Stephen Buay was the cause of the conspiracy since he was secretly assigned by Commander Salva Kiir to cause confusion and tension in Western Upper Nile so that he could stage a coup against the Commander-in-chief of the SPLA.

Then, the Commander-in-Chief asked Peter Gatdet, "who do you think can act in your absence till your case is resolved with your commanders"? Peter Gatdet refused to recommend anyone as he wanted himself to come down from Nairobi where he was and go directly to the field to command his forces. John Garang then suggested another name of Commander Stephen Duol Chuol to him to act in his absence until the case between him and his commanders was resolved.

Peter Gatdet finally agreed with the Commander-in-Chief that Commander Stephen Duol Chuol act in his absence and immediately Stephen Duol Chuol was rushed to the area to control and command the forces of Western Upper Nile in acting capacity. He was warmly received by Alternate commanders (A/Cdrs) Mathew Puljang Top, Tito Biel Wic and Michael Kolchara Nyang who were on the ground with the forces.

Peter Gatdet instead, decided to report to Sudan Embassy when he failed to establish contact with his forces on the ground to arrange for him a plane to take him to Khartoum. He made this arrangement since he was misinformed that Stephen Buay was already in charge of the command and he would be killed if he went to the field. Due to this confusion, Peter Gatdet decided to defect and rejoined General Paulino Matip who welcomed him back and appointed him as Chief of Operations for his forces.

51

On hearing news of Peter Gatdet's defection, the Commander-in-Chief convened an urgent meeting with commanders that had worked under Peter Gatdet, and he asked Taban Deng Gai to attend. The Commander -in-Chief ordered all the commanders to work hand in hand with Commander Stephen Duol Chuol as the acting commander of Western Upper Nile. After the meeting, John Garang' made the following deployment:

1. Commander Stephen Duol Chuol - Commander of Western Upper Nile.
2. Commander Philip Bipean Machar - Deputy Commander of Western Upper Nile.
3. Commander Stephen Buay Rolnyang - Chief Relief Coordinator for Upper Nile to base in Lokichogio.
4. Commander John Jok Nhial - Chief Administration for Western Upper Nile.
5. Commander Samuel Gai Yirchak - Commander of Mayom Operations.
6. Commander Karlo Kuol Ruac - Commander of Bentiu Operations.
7. Commander Michael Chiangjiek - Chief of Military Intelligence for Western Upper Nile command.
8. Commander John Puoljor Wicyoak - Western Upper Nile headquarters.
9. Commander James Nhial Wathkah - Western Upper Nile headquarters.

The Commander -in-Chief gave instructions to the commanders to report to the ground in Western Upper Nile and take charge and ensure Peter Gatdet did not control the forces. The commanders were moved from New Site by plane and landed at Riah and immediately came face to face with resistance on arrival at the airstrip. They were not allowed to disembark from the plane by the three alternate commanders (A/Cdrs) Mathew Puljang Top, Tito Biel Wic and Michael Kolchara Nyang who had arrested them earlier before Peter Gatdet defected and sent them to New Site.

Commander Stephen Duol Chuol pleaded with those of Mathew Puljang Top to allow the commanders to land and go to their homes until the issue would be resolved. His plea was accepted by Mathew Puljang Top, Tito Biel Wic and Michael Kolchara Nyang. They allowed the commanders to go on condition that they report to their respective homes and not give any directive on any command activity in Western Upper Nile.

After this tension, Stephen Duol Chuol recommended the three A/Cdrs Mathew Puljang Top, Tito Biel Wic, and Michael Kolchara to the Commander-in-Chief to be promoted to the rank of full commander so that they could remain in their respective commands. The Commander-in-Chief immediately approved and promoted the three A/Commanders to the rank of full Commanders, and they maintained their command status quo in Western upper Nile.

SPLA Western Upper Nile Command, 2003-2006

I went to Lokichogio, Kenya to assume my new position as Chief Relief Coordinator for Upper Nile region, knowing that my deployment to replace Peter Gatdet was sabotaged and undermined by Commander Taban Deng Gai who convinced the Commander-in-Chief to assign me to Lokichogio, Kenya to replace Michael Chiangjiek Gey who was working there as Upper Nile Relief Coordinator. I did not stay long in Lokichogio, as I wanted to visit my family whom I had not seen for some time. I requested for permission from Commander Salva Kiir Mayardit, the Deputy Chairman and Commander-in-Chief of the SPLM/A to allow me to go to Bahr El Ghazal to visit my family. He granted me permission and I flew to Lietnhom in Gogrial County, where my family was, and spent one month there on holiday, ostensibly to catch up with my family members after a long absence.

While still on holiday, I heard that Dr Riek Machar and Taban Deng Gai were going to Western Upper Nile on a tour. I asked for permission from Commander Kiir to go to Western Upper Nile to meet with them and he granted me another permission to go to Western Upper Nile. I walked on foot together with my bodyguards from Lietnhom to Western Upper Nile and met Riek Machar and Taban Deng Gai at a place called Lare area, and together we moved to Mayenjur. While at Mayenjur, Riek Machar ordered for Abiemnom area to be carved out from greater Mayom as an independent county and from there we proceeded back to Lare and went to Riah where Riek Machar called all commanders in Bul area for a meeting. There, Riek Machar asked Commanders Mathew Puljang and Tito Biel Wic to choose between Commanders Philip Bipean Machar and Stephen Buay Rolnyang who should remain as the overall commander in Bul area.

All commanders agreed and recommended me to be their overall commander in Bul area. Thereafter, Riek Machar and Taban Deng Gai left me in Riah with commanders Mathew Puljang and Tito Biel Wic as they proceeded and arrived in Leer area. On arrival, Riek Machar changed his mind, and sent a message to all units appointing me as the Commissioner of Mayom County and appointed Commander John Jok Nhial, as the Commissioner of Rubkona County. I felt unhappy when I received the message, because I had not expressed interest to be the Commissioner of Mayom County. Secondly, Riek Machar had no authority to appoint any SPLA senior officer to a position that was reserved for a civilian without the knowledge of the Chairman and Commander-in-Chief of the SPLM/A. I wrote back and informed him that I was not willing and not interested in taking up the new appointment as the Commissioner of Mayom County.

I was then reassigned as Chief of Operations for Western Upper Nile command under the direct command of Commander Stephen Duol Chuol and when Commander James Hoth Mai was appointed as overall Commander of Upper Nile in 2003, he appointed me as Commander of Western Upper Nile, deputized by Commander John Jok Nhial.

Another deployment was made after three months where Commander Peter Bol Kong was redeployed as Commander of Western Upper Nile, and I became the deputy Commander of Western Upper Nile command. I worked with Commander Peter Bol Kong for one year as his deputy before he was transferred from Western Upper Nile to Lou Nuer area. After his transfer, I was appointed for the second time, as Commander of Western Upper Nile and from there I assumed the command of Western Upper Nile from 2004 to 2006, after which I was transferred from Western Upper Nile to Kapoeta area.

> Note: *The Nasir coup was not tribal; it was a leadership and power struggle within the SPLM/A. As a result of command failure on both sides, the leadership of the two factions could not control their rogue and unruly commanders and militias who took the law into their own hands targeting tribal members of either side on the ethnic lines.*

SPLM/A Members of High Command

Figure 7: From left to the right is William Nyuon Bany, Dr John Garang, Salva Kiir Mayardit and Kuol Manyang' Juuk.

Figure 8: John Garang (left), Kerubino and William Nyuon (right)

Figure 9: Kerubino Kuanyin Bol (left), Arok Thon Arok (Middle)
William Nyuon Bany (Right).

Figure 10: Dr Lam Akol Ajawin

Figure 11. Dr Riek Machar and his wife

Chapter 4

Secret Behind Dinka-Nuer Onslaught

The Dinka, also known as Jieng, and Nuer also called Naath are the two largest ethnic groups in South Sudan forming the first subdivision of the Nilotic group. A second subdivisioyn comprises the Shilluk and various peoples who speak languages similar to it such as Luo, Anuak, and Lango. Dinka and Nuer are tall, long-limbed, and physically and culturally alike. Their languages and customs are too similar though the history of their divergence is unknown, according to an old myth, Nuer and Dinka were sons of one of the great-great-grand ancestors from time immemorial who was blind and promised to divide his old cow and its calf between Dinka and Nuer. He gave the cow to Dinka and gave its calf to the Nuer, but Dinka was unhappy with the distribution because the cow was old, and he wanted to be given a calf. When they went away, the Dinka came back at night to their father by imitating the voice of the Nuer, then the father thought that he was the Nuer talking to him and gave him the calf when the Nuer came to take his calf, the father found out that he had been tricked by the Dinka to take the calf of the Nuer.

The father got angry and told the Nuer to go after the Dinka to avenge his calf by raiding the Dinka whenever the calf becomes an adult and calving other cows, from henceforth, the Nuer raid the Dinka cattle by war and the Dinka steal Nuer cattle by robbery. The Nuer regard Dinka as thieves and Dinka regards the Nuer as aggressors. The raiding of the Dinka cattle by the Nuer is conceived to be a normal situation and duty, for they have a myth that explains and justifies their raiding of Dinka cattle. The word Nuer is a Dinka origin.

Dinka and Nuer had no form of government from time immemorial, they lived in anarchy and lacked formal laws. Their political development began during the emergence of their prophets (Sky-gods), spear masters, and leopard skin chiefs who had ritual powers to deal with social life and nature, including the power to bless and curse. Dinka and Nuer are herdsmen and the only labour they delight in is the care of cattle. The cattle are their dearest possession, and they gladly risk their lives to defend their herds or to pillage the cattle of their neighbours. Most of their activities concern cattle. The attitude and relationship of Dinka and Nuer towards neighbours are influenced by their love of cattle and their desire to acquire them. Their wars against each other have been directed to the seizure of cattle and control of pastures and water points.

The two ethnic groups have had long conflicts over grazing land, cattle, and water with the Sudanese nomadic Messiriya Arabs (Baggara ethnic groupings of Arab tribes) in Darfur and Western Kordofan who have been armed by the Sudan government to fight them and expand land occupation towards their territory. The Nuer and Dinka confronted the Messiriya Arab tribes with traditional weapons spears, shields, and clubs and as a result, much of their territories have been occupied by the Messiriya Arab tribes.

This occupation forced most of the Nuer and Dinka youths to join the Sudanese second civil war to acquire arms to defend their cattle and land. They, however, ended up being trained and oriented toward the liberation struggle against the Arab cliques who ruled Sudan since the attainment of independence on January 1, 1956.

Dinka and Nuer carried on the cattle disputes with them to the Sudan People Liberation Movement and Army (SPLM/A) which was formed in 1983, with the ideological objectives of the liberation of the whole of Sudan to create a New Sudan. The idea, however, never saw the light of day as with time, members of the SPLM/A political-military high command began to doubt the realization of the ideology of fighting for the liberation of the whole of Sudan. There was a need for them to review the manifesto of the SPLM/A to change the ideology from fighting for the whole of Sudan, which they thought, would take time and resources and wanted a change of plan and ideology to fight only for the liberation of Southern Sudan including Abyei, Nuba Mountain, and Southern Blue Nile. Members of the SPLM/A political-military high command asked Dr. John Garang to convene an urgent high command meeting where they would discuss and review the SPLM/A manifesto, but John Garang who had already learned about the plan did not convene the meeting and instead ordered a mass mobilization campaign of the SPLA forces to converge around Juba to attack and capture Juba.

Two high command members in the Nasir faction Commander Riek Machar Teny Dhurgon and Commander Lam Akol Ajawin refused to contribute forces to go to Juba. They instead decided to act without waiting for the rest of the members of the SPLM/A political-military high command who were sharing the same ideas with SPLA senior officers who also wanted to overthrow John Garang. However, on August 28, 1991, they became impatient and sent a radio message to all the SPLA units that they had taken over the leadership of the SPLM/A and accused John Garang of dictatorship and running the movement alone. They had promised to promote and advocate for human rights within the movement and called for the self-determination (**New Ideology**).

The move by the Nasir faction was received negatively by the rest of the SPLM/A high command members who were having the same idea but were pre-empted by their junior commanders in Nasir. As a result, they came up with a plan to stop the Nasir group in its track by issuing covert orders to target Nuer SPLA soldiers and innocent civilians in the areas controlled by the SPLM/A mainstream known as the Torit faction. The aim was that when the Nuer heard about this development, they would also target Dinka in Nuer-controlled areas, which was a great plan for them to prevent the rest of the Dinka SPLA soldiers from supporting the Nasir coup.

On the same development, some elements from the Nasir faction reacted heinously by targeting Dinka in the Nuer-controlled areas and therefore, the conflict changed from being a leadership takeover to a tribal dimension which was exactly planned by the SPLM/A violent extremists. The plan to fail the Nasir coup had successfully worked out by the fact that all Nuer survivors in the areas controlled by the SPLM/A mainstream managed to desert to the areas under the control of the Nasir faction and all Dinka survivors in the areas controlled by the Nasir faction deserted to the areas controlled by the SPLM/A mainstream.

Both Nasir and Torit factions carried out massive campaigns to attack areas controlled by either side including innocent civilians, with the Nasir faction on the offensive using armed Nuer youth known as White Army from greater Lou and Fangak to attack greater Bor in Jonglei and Nuer armed youth from Western Upper Nile (Bentiu) to attack greater Warrap in Bhar El Ghazal which involved cattle raiding and wanton vandalism.

The secret plan was that we target the Nuer so that the Nasir faction would react negatively by targeting the Dinka so that there would be no Dinka who would go over to the side of the Nasir faction. This plan worked out as envisaged as no SPLA Dinka soldiers dared to join the Nasir faction when it was targeting the Dinka. If the Nasir faction had not reacted negatively by targeting the Dinka SPLA soldiers in the Nasir faction-controlled areas and tended to behave as revolutionaries and nationalists, most of the SPLA forces if not all, would have abandoned Torit faction under John Garang and joined the Nasir faction and there would not be enough forces to remain with the SPLM/A mainstream.

In 2013, some senior members of the SPLM led by former Vice President Riek Machar Teny Dhurgon decided to challenge the chairperson of the SPLM, in the person of President Salva Kiir Mayardit to give chance to one of the SPLM senior members to be an SPLM flag bearer in the General Elections which were supposed to be conducted in 2015. President Kiir was shaken by the move and almost lost the political ground within the SPLM party, but he was lucky as he was rescued by the planners of the 1991 anti-Nasir faction plot, who came up and repeated the same idea by targeting the Nuer in Juba so that the Nuer would do the same as they did in 1991, and the SPLM leadership struggle would be turned into a tribal conflict that would save President Kiir from being removed from the chairmanship of the SPLM and use the ethnic conflict to consolidate himself in absolute power. As a result, over 20,000 Nuer ethnic people were killed in Juba on tribal lines. "The deaths of Nuer and Dinka will never end if President Kiir and 1st Vice President Riek Machar are in power. They will always use Dinka and Nuer to kill themselves whenever their leadership positions are under threat and the rest of 62 tribes of South Sudanese will be the grass to suffer."

> Note: *The dominant perception in public discourse around the world is that hostility between Dinka and Nuer is tribal, but the reality is far from it. It is a tool being used by political entrepreneurs not only to get and stay in power but also to advance other parochial interests, patronage, and cronyism*

Chapter 5

Leadership Failure in South Sudan

The January 9, 2005, signing of the Comprehensive Peace Agreement between the Government of Sudan and the Sudan People's Liberation Movement/Army (SPLM/A) was a historic moment of great opportunity for the country and one which all its people expected to offer them a path toward recovery, restoration of their rights and to a solid and long-lasting peace. John Garang de Mabior, the founding father of SPLM/A signed the agreement with the Sudan government under Omar Hassan El Bashir, with political and diplomatic support from the United States of America, United Kingdom, and Norway (Troika partners), and other member countries of the European Union, United Nations, and African Union (AU). The transitional Government of Southern Sudan (GOSS) set up in 2005 led the people of South Sudan through a referendum and subsequent independence of the country in 2011. It gave the people very high expectations and hope not just that the guns, which had affected many different areas of life would stop firing but that peace, prosperity, and economic stability could return to the Republic of South Sudan.

The CPA ended the longest Sudanese Civil war that dated back to the mid-1950s, when Southern insurgents took up arms against the Islamist government in the Sudan who were oppressing the Southerners, the war which resulted in the deaths of more than two million people, just between 1983 and 2005.

Unfortunately, John Garang was killed in a Uganda helicopter crash on July 30, 2005, and because of the tragic death, President Salva Kiir who was an SPLA battlefield commander with little knowledge of the democratic governance and rule of law was shortly installed to lead South Sudan, home to over 12 million people drawn from 64 tribes or ethnic groups with diverse cultures, religion, and languages. With the death of John Garang De-Mabior, SPLM lost direction and failed to uplift the country and the people from years of neglect, endemic poverty, lack of service delivery, and being the marginalized fringe of Sudan. The SPLM leaders missed the opportunity to consolidate peace and national unity and to seed a much-needed democratic transformation of South Sudan.

Although GOSS was well endowed with plenty of South Sudanese, regional, and international goodwill, donor funds, and oil proceeds and well positioned to quickly set up a functional system to deliver on the expectations of the people, it failed to deliver. Kiir appointed his long-time rival Dr. Riek Machar as his 1st vice President. Both leaders failed to meet the expectations of the people of South Sudan who were fighting for freedom, justice, and equality because they felt they were being oppressed by the Arabs. However, it has turned out that the oppression of Salva Kiir and Riek Machar is worse than the oppression of the Arabs.
The South Sudanese people expected the two leaders to guide them toward a genuine democratic system of governance in the country.

They, however, failed to establish a viable state and to build vital institutions right from the independence of the Republic of South Sudan. The only institutions that exist in South Sudan today are the factional and militia army (SSPDF) and the national security forces, which are predatory. After being in the presidency for two years, Salva Kiir and Riek Machar revealed their own true colours of being dictators and plunged the country into a civil war, which some South Sudanese political and armed groups argue has been motivated, by ethnic hatred and ethnic conflict. I really disagree with this argument and conceptual thinking. In fact, the root causes of the conflict can be traced back to the SPLM/A-Anyanya II power struggle and leadership failure that has been explained thoroughly in previous chapters. The peace agreement was signed following pressure from the international community that wanted to see that human rights abuses, especially on the South Sudanese, were stopped.

65

Under the CPA, a referendum was to be conducted to enable the South Sudanese people to make one last decision about whether to remain in a united Sudan or go separate ways and form the government of South Sudan that was to be ruled by South Sudanese themselves. It was also agreed that the two standing armies form a joint integrated unit to be drawn from both parties for a six-year period, after which a referendum was to be held for the people of South Sudan to decide their fate. The idea of a referendum was meant to see them remain in Sudan as a country or allow them to break away and go as a separate sovereign state.

On July 8, 2005, John Garang', the leader of the SPLM/A, set foot in Khartoum for the first time in more than two decades following the civil war that was mainly fought in the bushes of South Sudan. He was sworn in as the First Vice President of the Republic of Sudan and at the same time as the President of the Government of South Sudan (GOSS). Salva Kiir Mayardit became his Deputy in the government of South Sudan.

Unfortunately, Garang did not enjoy the dividends of the long struggle and the peace he signed with the Sudanese government after decades of misunderstanding that led to war. He died in a Ugandan helicopter crash on July 30, 2005, as he was returning to his base at New Site, following a meeting in Kampala with Ugandan President Yoweri Museveni. Upon this sad development, the SPLM leadership met at New Site and nominated Salva Kiir Mayardit to replace John Garang as the First Vice President of the Republic of Sudan and the President of the Government of South Sudan in accordance with the political and military protocol of the SPLM/A. Riek Machar was selected as the 1st Vice President of the government of South Sudan (GOSS) and James Wani Igga as the 2nd Vice President.

After all these confirmations, on August 11, 2005, Salva Kiir went to Khartoum, Sudan where he was sworn in as the 1st Vice President of the Republic of Sudan and the President of the government of South Sudan. President Kiir negotiated a pact aimed at bringing all other armed groups that were fighting the SPLM/A alongside the Sudan armed forces on board and managed to convince General Paulino Matip Nhial to join the government of South Sudan. As a result, in 2007, General Paulino Matip Nhial and his group joined the government of South Sudan under what was called Juba Declaration Agreement. The forces of the SSDF were integrated into the SPLA, and General Paulino Matip Nhial was appointed Deputy Commander-in-Chief of the SPLA, a position that was not in existence before, but was created to accommodate him.

Power Struggle in the SPLM, 2013

The third power struggle within the top leadership of the SPLM ruling party started in 2013 when Salva Kiir Mayardit accused his Vice President Riek Machar Teny Dhurgon of attempting to oust him from power. The move spilled over into a military confrontation between the forces loyal to Salva Kiir Mayardit and Riek Machar, plunging the whole country into civil war. As a result of total leadership failure, the unruly and rogue militias known as Mathiang-Anyor and some other uncivilized Dinka men in Juba took the law into their own hands by targeting Nuer in Juba. The conflict took a tribal dimension and spread fast across the country causing serious destruction with civilians of either Dinka or Nuer being targeted on ethnic lines and displacement of more than two million people.

The commencement of a full-scale civil war resulted in the death of nearly 400,000, and the displacement of more than two million people. Ugandan President, Yoweri Museveni sent troops to Juba to fight alongside the SPLA and to protect the seat of government since the SPLA had split along tribal lines. Some Ugandan soldiers were deployed in Bor town to defend it and protect the civilians and their properties. The conflict spread in many towns in Upper Nile where people's properties were set on fire and razed to the ground. When Dr Machar's faction failed to seize power, his group took the new name as the Sudan People Liberation Movement/Army in the

Opposition (SPLM/A-IO).

Several ceasefires were signed and subsequently got violated by both sides during this period until a final peace agreement was signed in Addis Ababa under the auspices of the Inter-Governmental Authority on Development (IGAD) following intense pressure and threats of enforcing sanctions against the two protagonists by the United Nations. After the agreement had been signed, Machar was installed as the 1st Vice President of the Republic of South Sudan, and he returned to the capital Juba in 2016 to take the oath of office.

However, before the ink of the agreement could dry, an intensive fight erupted again between the factions at the Presidential Palace, J1, in Juba. Referred to as the J-1 dogfight, Machar was forced to flee the country again and he crossed over to DR Congo with his forces. Taban Deng Gai was appointed 1st Vice President, replacing him and representing SPLM/A –IO. The lack of leadership qualities in the two leaders, President Kiir and his Vice President Riek Machar continued to breed tribal nationalism and hatred between Dinka and Nuer ethnic groups. They failed to control their tribal militias and instead armed them to fight for them as they continued their leadership struggle.

Revitalized Peace Agreement

In 2018, the peace that was violated in 2016, as a result of the J-1 presidential palace fight between forces loyal to President Kiir and his former Vice President Riek Machar, was revitalized under the auspices of the Inter-governmental Authority on Development (IGAD) which led to the signing of the final peace agreement in Khartoum known as the revitalized agreement on the resolution of the conflict in the Republic of South Sudan that reinstated Riek Machar as 1st vice president and this time around with other more vice presidents which resulted in the formation of the fragile and embattled revitalized transitional government of national unity.

It was agreed controversially to hold elections on December 22,

2022, but this failed because the implementation of the peace agreement was deliberately delayed and undermined by the SPLM-In government. The parties to the revitalized Transitional Government of National Unity (RTGONU) met and charted the way forward by extending its term under what they called the new **roadmap** for the revitalized transitional government of national unity. The new roadmap commenced on February 23, 2023, with the aim of implementing the rest of the issues or provisions of the agreement which were not implemented and to pave the way for the General Elections that will take place on December 24, 2024.

The extension of the RTGONU was a local arrangement by the parties to the agreement which lacked international and regional back up and as a result, President Kiir who saw the legality of the RTGONU end on February 23, 2023, went in and dismantled the roadmap agreement. He issued the republican order on March 3, 2023, dismissing the Minister for Defence and Veteran Affairs, Angelina Teny, the wife of 1st Vice President Riek Machar Teny Dhurgon who was appointed to the position on the ticket of the SPLM-IO, the position she held since the signing of the peace agreement.

President Kiir did not stop there; he issued another order swapping the Defence ministry occupied by the SPLM-IO with the Interior ministry occupied by the SPLM-IG without any further consultation with the leadership of the SPLM-IO. The SPLM-IO did not take the matter lightly and described the move as a unilateral decision made by the President.

They called for the SPLM-IO Political Bureau meeting where they came out with a resolution of rejecting the order of the President by removing Angelina Teny Dhurgon from the position of Minster for Defence and swapping it with the Ministry of Interior which belonged to SPLM-IG. The SPLM-IO Political Bureau asked President Kiir to rescind his republic order. Transitional Government of National Unity is essentially the sixth transitional term of the same

69

Government in South Sudan. The RTGONU has been led from 2005 to date by the same crop of SPLM leaders who have miserably failed the country and failed to deliver on the mandate of the Agreement on the Resolution of the Conflict in the Republic of South Sudan (ARCSS), and subsequent revitalized version of the same agreement (R-ARCSS) despite the protestation by those advocating continuity of the same. The series of transitional governments post the signing of the CPA are as detailed below:

1. Government of Southern Sudan (GOSS), 2005 – 2011.
2. Transitional Government of South Sudan, 2011 – 2015.
3. Transitional Government of National Unity (TGONU), April 2016 - July 2016.
4. Second Transitional Government of National Unity (TGONU), 2016 - 2018 (led by Salva Kiir and Taban Deng Gai instead of Riek Machar as provided for in ARCSS).
5. Revitalized Transitional Government of National Unity, 2018 - 2021, extended to February 2022 (R-ARCSS).
6. If allowed to stand, a new unilateral or self-revitalized regime of President Salva Kiir, Feb 2023 - Feb 2024 (R2-ARCSS).

The extension of the RTGONU has been rejected by a group of concerned and like-minded stakeholders called the National Consensus Forum (NCF) who strongly believe in and advocate for a new political dispensation for South Sudan which is urgently needed to move the country forward in the interest of the people and away from the current threshold of abject failure in South Sudan, beyond which the country is sure to plunge into anarchy and total collapse.

The new political dispensation is to be based on an all-inclusive and people-centric roundtable conference (RTC) that shall be endorsed by various Political Parties, Civil Society Organizations, Faith-based Groups, Academic and Professionals, Women, and Youth groups that are parties to the NCF. In a meeting held on September 5, 2022, the like-minded stakeholders for a new political dispensation for

South Sudan resolved to develop a National Consensus Statement (NCS) that redefines the problem in South Sudan and outlines viable options to resolve the crisis. The group also agreed to work together as a broad-based multi-stakeholder forum that serves as an open platform for South Sudanese to dialogue candidly and freely to find durable solutions to the root causes of the conflicts that have afflicted South Sudan over decades. These include, particularly, internecine ethnic rivalries and struggle for domination that transcends South Sudan's first liberation movement (Anya-Nya war), the period of Southern Sudan's regional government (1972-1983) following the Addis-Ababa peace agreement, and the SPLM/A liberation era and GOSS following the CPA and independent South Sudan from 2011 to date.

The NCF invited all South Sudanese opposition groups, organized political and civic groups, women and youth organizations, academic and professional groups, faith-based entities at home and abroad, and friends of the people of South Sudan to rally and support the call for a new political dispensation for South Sudan.

The NCF members include:
1. The Non-Signatory South Sudan Opposition Groups (NSSOG) consisting of the National Democratic Movement-Patriotic Front (NDM-PF), National Salvation Front (NSF a.k.a. NAS), Real-SPLM, South Sudan National Movement for Change (SSN-MC), South Sudan United Front and Army (SSUF/A), and United Democratic Revolutionary Movement/Army (UDRM/A). Others are United Peoples' Democratic Party (UPDP), Red Card Movement (RCM), South Sudan People's Movement & Army (SSPM/A), National Peoples' Movement (NPM), South Sudan Steps Towards Peace & Democracy (STEPS), and South Sudan Peace Rally.
2. Civil Society groups which include Peoples' Coalition for Civil Action (PCCA), Cush Organization for Development & Advocacy (CODA), and South Sudanese Business Community in

71

Uganda (SSBCU).
3. Women and Youth groups: Strive Africa Action (SAA) and Ny-aeden.
4. Academics and Professionals.
5. Faith-based organizations: Redemption International Ministries, Christian United for South Sudan, and Agape International.

After staying in power for 16 years, and for his own political survival, Kiir has revealed his true colour by indulging in a dictatorial and kleptocratic tendency and has created his own private paramilitary security apparatus to intimidate, torture, execute people, lynch political dissidents for crimes they did not commit. His reign has also witnessed rampant corruption involving embezzlement or misappropriation or stealing of state resources and millions of dollars from the United States in form of grants and donations for the fiscal year and other humanitarian assistance for over 12 million poor people of South Sudan largely depend on the foreign aid.
Kiir has failed to resolve the proliferation of firearms, which is the cause of rampant inter–communal violence, revenge killing, rape, and cattle raiding in the countryside termed by the UN as sub-national violence, which has engulfed the whole country. The poor governance, mismanagement of oil revenues, and rampant corruption in the country have contributed to increased political and economic instability in the country. The RTGONU has been extended for more than two years starting from February 23, 2023, to December 2024, when General Elections will take place.

Sudan Population and Housing Census

The Sudan 5th Population and housing census was conducted in 2008 before the separation of South Sudan from the rest of Sudan in January 2011, with some areas not covered, especially in South Sudan, but just estimated through traditional authorities (chiefs).

The following are the details of the population and housing census that has been enumerated in Sudan since 1955.

S/N	Year	Total
1.	1955/1956	10.3 million
2.	1973	14.8 million
3.	1983	20.6 million
4.	1993	25.9 million
5.	2008	32.07 million

After the comprehensive peace agreement, the Sudan 5th population, and housing census were conducted in 2008, it was the most important census in the history of Sudan though it was not complete, especially in South Sudan due to insecurity in some areas. In addition, the refugees who had scattered in the region were not properly enumerated and there was no other population census conducted so far in South Sudan after the independence.

The SPLM-led government went ahead and conducted secret population estimates in 2021 through the National Bureau of Statistics, which they released on 6 April 2023, which is regarded by the people of South Sudan as an attempt to rig the upcoming elections in advance. There is a great disparity between the SPLM population estimates and the Sudan 5th population census that was last conducted in 2008, before the separation of South Sudan from the rest of Sudan and the subsequent independence of the Republic of South Sudan. The last Sudan 5th population census is hereby contrasted with the SPLM population estimates as arranged in the table below: -

S/N	State	Sudan 5th Population Census 2008	SPLM Population Estimates 2021
1.	Upper Nile state	964,353	790, 147
2.	Jonglei state	1,358,602	791, 105
3.	Unity State	585,801	892, 780
4.	Western Bhar El Ghazal state	333,431	562, 555
5.	Northern Bhar El Ghazal state	720,898	1, 924, 342
6.	Warrap state	972,928	2, 639,484
7.	Lakes State	695,730	1, 265, 473
8.	Western Equatoria state	619, 029	663, 233
9.	Central Equatoria state	1, 103,592	1,324, 521
10.	Eastern Equatoria state	906,126	981, 902
11.	Pibor Administrative area	-	240, 102
12.	Ruweng Administrative area	-	234, 416
13.	Abyei Administrative area	-	133, 958
14.	G/total	8, 260, 490	12, 444,018

The disparity shows that Warrap alone has more people than the three states of greater Upper Nile, Jonglei, Upper Nile, and Unity states combined, which indicates that the votes in Bhar El Ghazal alone can give a win to whoever can steal the votes, in both presidential and parliamentarian elections. Secondly, if you have a proper look at the comparison between the SPLM population estimates with the Sudan 5th population census, it means that 434,724 people have died or been killed in greater Upper Nile over the last 15 years from the Sudan 5th population census, and it also suggests that nobody was born over the last 15 years in greater Upper Nile, and **1,666, 556**, people were born in Warrap and also suggests that nobody died or killed over the last 15 years in Warrap state.

Furthermore, the SPLM population estimates suggest that the remaining population of the greater Upper Nile will be extinct and the current population of Warrap will double in the next 15 years. These are malicious intentions designed by the SPLM to outsmart the people of South Sudan to rig the upcoming elections. These SPLM population estimates are fake and must be disregarded and cannot be used in the upcoming elections. The R-ARCSS says a new census and voters register will be conducted by the parties to the agreement before the elections could take place.

> Note: *While some people may see violence in South Sudan as all about "tribes", the reality on the ground shows poor leadership, corruption, and bad governance by elements whose only interests are to seize and retain power using their ethnic henchmen who are perpetuating the conflicts that have killed thousands and sent millions of South Sudanese to various refugee camps in the region.*

Chapter 6

The Effects of Dictatorship Rule and Leadership Failure

The Republic of South Sudan became the world's youngest nation and Africa's 54th country on July 9, 2011, after decades of bloody struggle with the government in Khartoum in which millions of people lost their lives. The day brought hope for a better future and the opportunity to build a united state founded on justice, equality, respect for human dignity, and advancement of human rights and fundamental freedoms. However, more than a decade later, the promise of South Sudan's independence turned into a tragedy.

The country remains affected by underdevelopment, mismanagement, corruption, inter-ethnic conflicts, and entrenched dictatorship. Dictators rise to power in several ways, one being the launching organization. If the organization is weak, dictators will try *to concentrate power and resources in their own hands and slowly transition into a personalist regime in which one powerful individual dominates the government apparatus and its instruments.* Dictators always strive to concentrate power into their own hands in order to increase their security at the top and because there are usually no institutionalized mechanisms for succession to hand over power to someone else, they become very paranoid.

The dictator does everything he can to eliminate potential rivals, by deliberately weakening the military to prevent coups which is known as coup-proofing. The military could be weakened by not having training, not having access to weapons, or creating a parallel military organization to offset the power of the traditional military. Furthermore, the dictator deliberately weakens his own political party by ousting those who have the most expertise or those who have the potential to possibly challenge him. The other thing that the personalist dictator may want to weaken is the legislative branch and the judicial branch. The courts are deliberately weakened by selecting people who are going to listen to what he has to say, or he can threaten their lives or their jobs if they do not make decisions that he agrees with. He can also weaken the legislature by literally directing who gets to be in a particular office, purging individuals that he finds to be disruptive or a threat to him. The dictator may weaken the bureaucracy by playing musical chairs with the individuals who are already part of the system, by putting them in positions for just a couple of years and then transferring them to another position. He can also do this by firing and hiring them again, making them feel completely insecure and creating a very chaotic environment.

He will weaken the bureaucracy so much that it barely even functions and the purpose of this is that he does not want any group of people to be able to have any kind of expertise or ability to challenge him. So, the dictator prefers, in the personalist case, where you have just one person ruling.

As a result, the regime is completely de-institutionalized and completely personalized under the power of this one leader. When the dictator sees his life in power as possibly short, he believes that he needs to hoard as much as possible and steal from the state. Since there are few checks on his power, he believes that he is above the law. He sees life in power as very fleeting. He thinks he may even have a violent exit because he sees his time horizon is short, he decides that while he is in power, he is going to hoard as much as possible in large volumes.

The other aspect that leads a dictator to more kleptocratic behaviour is that there are absolutely no checks on the dictator's power to ensure or prevent him from stealing. So, he can steal large amounts of money, and no one says anything about it, and then that makes things even worse because the culture of corruption develops where they allow these large amounts of wealth to be stolen with nothing being said about it and because institutions are extremely weak, there is no transitioning of power from one leader to the next. The mode of exit, when it comes to that, is also extremely violent and the dictator lives in fear.

The very question that is always ringing in the ears of a dictator is what life will be like after leaving power. **The answer of course is either jail, exile, or death.** This makes it much less likely that the dictator will step down. He may need to die in power and surrounds himself with sycophants who report to him falsehoods and lies and tell him what he wants to hear and so this can shape the personality to make a dictator even more narcissistic and delusional than he may already be. The sycophants tell him that they are the best and greatest, they need to be in power forever, that anyone who tries to challenge them would die, and that anyone of their enemies is weak. This makes it difficult for a dictator to step down and instead cling to power until the very end, seeing no life for himself after being in power and only choosing to die in office. The following are different methods or approaches to be used to get rid of a dictator:

1. He can be deposed in a coup by the junta.
2. He can be forced to resign by his elites.
3. He can be assassinated.
4. He can be defeated by violent revolution (Rebellion)
5. He can be forced to resign by non-violent protests.
6. He can be forced to relinquish power through international intervention (**which takes time**)

This is exactly what is happening in the Republic of South Sudan. President Salva Kiir and his Vice President, Dr. Riak Machar, focused their attention on the control of power and militarization of their ethnic support base to pave the way for a grip on the leadership of South Sudan. Dictatorship took root in South Sudan, and a kleptocratic patronage system was born and consolidated when Dr. John Garang died in a Ugandan helicopter crash and Salva Kiir was nominated by the SPLM leadership to replace him.

On Salva Kiir's takeover, insecurity and lawlessness were immediately unleashed as tools of repression to obscure the plunder and misappropriation of national resources. Abuse of human rights and dishonour of agreements cemented a perennial rule of a self-mutating transitional government of a crop of the same leaders buoyed by power-sharing agreements serving the interests of the SPLM political and military elites, including the Jieng Council of Elders (JCE), NCP, and the current business cartels as part of these self-serving groups and not the South Sudanese.

The SPLM was one of the greatest movements in Africa before it was infiltrated and eventually hijacked by the members of the National Congress Party (NCP). President Kiir has weakened the SPLM Party by ousting those who have the most expertise or those who have the potential to possibly challenge him, and as a result, the SPLM has disintegrated into 7 SPLM parties namely: SPLM in government, Yei faction led by President Salva Kiir, SPLM- in Opposition-1 led by Riek Machar Teny, SPLM-in Opposition-2 led by Taban Deng Gai, Real-SPLM Opposition faction led by Pagan Amum Okiech, SPLM-FDS (Former Detainees) Opposition faction led by Mama Rebecca Nyandeng, the widow of Late Dr. John Garang, SPLM-DC, Opposition faction recently changed to National Democratic Movement (NDM) led by Dr. Lam Akol Ajawin, and SPLM-NCP led by Tut Keaw Gatluak.

In this case, the ruling SPLM party becomes weak, and President Kiir takes advantage of the situation to concentrate power into his own hands working closely with the NCP members leaving the real SPLM members in the cold because he is so consumed with maintaining power and there are no institutionalized mechanisms for succession to hand over power to someone else, he becomes very paranoid and does everything he can to eliminate potential rivals, by torturing and intimidating his critics and political dissidents. The neglected, intimidated, and traumatized people struggled for their survival as the international community watched with little attention paid to the ethnic, and regional power dynamics, and tribal intrigues that underscores the suffering and exploitation of the people. The administration of President Kiir does not differentiate between private and public benefits. The narrow elites and inner circle of cronies have captured the wealth of the nation which has impacted the economic growth and aggravated the economic hardship, and as a result, the national army and civil servants go for many months without pay causing extreme abject poverty and unemployment in the country. The national coffer and public funds/resources have been depleted, stolen, and looted to the core in broad daylight by some regime powerful officials who hold the reins of power to loot the public funds and resources with impunity.

According to research A three-year investigation into a loan deal between a local company and a regional bank by The Sentry, an investigative and policy organization, exposed how laws were broken in South Sudan. Under the arrangement, sanctions may have been breached, and powerful individuals were enabled to benefit from the manipulation of businesses worth hundreds of millions of dollars. The loan deal skirted legislation on oversight, transparency, and competition and facilitated off-book government spending, including supplies of fuel to the South Sudan army. It also perpetuated a damaging reliance on future oil production to finance current spending, mortgaging the future prosperity of the country and its citizens.

The investigation found that Trinity Energy spent millions of dollars on "facilitation" and "business acquisition" costs for the Afreximbank deal, including 18.7 million South Sudanese pounds (SSP) ($125,000) in payments to the government committee responsible for approving the deal. During the implementation of the trade finance deal, Trinity Energy changed millions of US dollars on the black market, paid fake invoices overseas to disguise the black-market exchange of hundreds of thousands of dollars, and engaged in behaviours indicative of tax fraud.

The South Sudanese government's actions were at odds with its own laws on procurement, competition, transparency, and petroleum revenue management. The absence of checks to government power in South Sudan opens the door to commercial dealings that are based on personal relationships and the exchange of benefits and favours. The result is that companies and the executive branch can conduct business in the absence of oversight or transparency and with scant regard to the rule of law, economic sustainability, and the well-being of the nation and its citizens. State officials and businesspeople operate in an environment of impunity that incentivizes rather than deters negligence, rent-seeking, and illegal activity.

Trinity Energy used the Afreximbank trade finance facility to supply diesel and gasoline worth millions of dollars to the South Sudanese army at a time when government forces were involved in ongoing civil conflict. The deal was not made public, thereby evading checks and balances on government spending, including parliamentary oversight. South Sudan's army has been accused of war crimes and human rights abuses. Trinity Energy used a local travel company called Moonlight Travel and Tours to exchange US dollars for South Sudanese pounds on the black market in Juba, generating millions of dollars in foreign exchange profits and breaking South Sudanese law.

Absence of Rule of Law in South Sudan

More than 10 years after independence, South Sudan is still faring poorly, beleaguered by ills attributed to negligence, poor leadership, and lack of vision and direction. The country appears to have spent the years since 2011 chasing shadows because it has failed or neglected to focus on the question of leadership. Leadership failure in a country is like a shadow, darkening many areas and misleading. The administration of President Kiir has shown through acts

81

of commission or omission that it lacks the ability to provide direction, build requisite national consensus, and motivation to keep the hopes of the people alive. Bad governance is always associated with high corruption which undermines the regime's legitimacy and fosters political inequality and economic inefficiencies which ultimately promote tribal nationalism and hegemony over others.

After independence, efforts to promote good governance, rule of law, and nation-building failed because top members of the SPLM ruling party who were expected by the citizens to promote them have instead been involved in a deadly power struggle, putting the country on the verge of collapse and disintegration. *Although South Sudan* is founded on justice, equality, respect for human dignity, and advancement of human rights, an evident *lack of respect for the law remains a long-standing problem* in the country. Rule of law means that no one is above the law and the law applies to everyone and even the government must follow the law of the country as set out in the Constitution.

However, in the Republic of South Sudan, there is no rule of law because of leadership failure and as a result, everybody is taking the law into their own hands. That is why there is rampant ethnic violence across the country, which the United Nations has termed as sub-national violence. In fact, they are national violence because they affect the whole nation, and the government is unable to arrest the situation because tribal nationalism and hegemony prevail over the crippled national security institutions that are unable to provide protection and security for the citizens and their properties. Rule of law and democracy are associated with less corruption. Government officials are supposed to use public office for the public good than for private gain. This is not the case in the Republic of South Sudan where there is low-level democracy that promotes corruption, and lack of freedom of expression and association.

The disaffected and dissident groups and individuals are always being subjected to physical and mental torture, intimidation, and mistreatment by the security forces with impunity. The leadership does not follow the democratic values that include freedom of assembly, association, property rights, inclusiveness and equality, citizenship, voting rights, freedom from unwarranted governmental deprivation of the right to life and liberty, and minority rights.

The absence of the rule of law in the Republic of South Sudan since 2011, is the main obstacle in the struggle for justice, freedom, and equity even as South Sudan is spiralling toward total collapse and eventual disintegration. It appears the government has failed to provide security to its citizens, a situation that has seen inter-communal violence spread widely across the country. These ethnic conflicts have been engineered and sponsored by the regime in Juba as part of its ethnocentric policy of divide and rule to remain in power for the rest of its life and they are likely to evolve into another genocide as happened in 2013.

The conflicts have deprived the people of South Sudan of basic services such as clean drinking water, food, education, and essential health services. The instability of political and socio-economic in the Republic of South Sudan is characterized by rampant sub-national violence. The regime is to blame for the circumstantial incidents such as organized crimes, ethnic violence, corruption, illegal drugs, the proliferation of firearms, economic sabotage, environmental degradation, effects of natural calamities, tribal militias, and factional armies across the country as explained below:

Unknown Gunmen

These are organized crimes that are orchestrated by the regime's intelligence agents to terrorize people, especially on the main roads leading to Juba City and in the major cities across the county so that it is blamed on the insurgents. The regime security agents are arming and collaborating with the criminals and directing them to eliminate whoever they want and call them the unknown gunmen while they are the ones committing all those crimes and atrocities in the name of unknown gunmen.

Twic Versus Abyei Communities

A minor incident that occurred at Anet market because of a misunderstanding between the Abyei administrative area and Twic County administration over the control and management of the market's income/resources that could have been resolved amicably by the local officials of both sides, has now taken the shape of communal conflict that has pitted Twic community against Abyei community. The conflict has claimed many lives and displaced thousands of people to North Sudan and nearby villages. This conflict between Abyei and Twic communities has been sponsored and fuelled on the watch of the national security in Juba by political officials on both sides who are supposed to arrest the situation as nationalists. Both sides are supplied with ammunition from the national security who are fuelling the conflict. Anet market is located inside Twic territory, but it is inhabited by Abyei people who came to settle there during the SPLA war when people of Abyei were displaced by the Arab nomads of Messiriya and they were moved by the then SPLA administration to be settled South of Kiir River. It is high time for the people of Abyei also to resettle back north of Kiir River and leave Twic people to occupy their place of origin South of Kiir River. Nevertheless, the regime is unable to settle the dispute between the two communities amicably by telling Abyei people to cross back to their area of origin because the regime has a hand in the inter-communal conflict between Abyei and Twic.

Bor Versus Communities of Eastern and Central Equatoria

The minor incidents that started with a misunderstanding between Bor cattle herders and some farmers in central and Eastern Equatoria that could have been handled by the national and local officials peacefully have now taken a tribal dimension that has pitted Bor people against communities of greater Equatoria, and in fact, the issue has been generalized and presented as a problem between Dinka as a tribe versus tribes of greater Equatoria, which was not the initial case. Government officials and politicians on both sides are instigating and fuelling the hatred and hostilities between the Bor cattle herders and communities of greater Equatoria regions.

This was witnessed in a video clip that went viral on social media on January 24, 2023, when Kajo-Keji armed youth took to the forest and shot indiscriminately at the cattle believed to belong to Bor cattle herders. They killed 400 heads of cattle and left them lying dead in the middle of the forest. As a result, Bor cattle herders retaliated by attacking Kajo-Keji youth, killing over 20 on February 2, 2023. This happened on the eve of the visit of His Holiness Pope Francis, Most Rev. Justin Welby, Archbishop of Canterbury, and Rt. Rev. Dr. Lain Greenshields, moderator of the General Assembly of the Church of Scotland paid a pilgrimage visit to South Sudan on February 3, after visiting DRC -Congo.

The world church leaders called on South Sudan's leaders to respect dissenting voices and address the country's ongoing human rights crisis and widespread impunity. Civilians in various parts of the country continued to face violent attacks. Since mid-2022, clashes between armed opposition groups and government forces and allied armed groups in Upper Nile and parts of Jonglei State where armed opposition groups are fighting for political and territorial control, have been accompanied by serious human rights abuses and the displacement of thousands of people.

Jonglei Versus Pibor

Incidents that started with the common cattle-raiding and theft among the tribes in the Jonglei region which was a result of the absence of the rule of law during the SPLA war and that could now be deterred by the current respective governments of Jonglei state and Pibor administrative area, has now taken tribal dimension that has pitted tribes of Jonglei against Murle people in Pibor administrative area. The conflict has been instigated and fuelled by their respective government officials and politicians working in the national and local governments. The conflict has claimed hundreds of lives and has seen thousands of cattle raided on both sides and children abducted. The rivalry between the Lou Nuer and the Murle remains one of the most violent communal conflicts in South Sudan despite numerous resolution attempts by the regime, the international community, and various Christian churches. Both communities share a long history of conflicts over resources, mainly the cattle raiding and abduction of children. However, the nature of cattle raiding has changed drastically in recent years with the proliferation of sophisticated weapons brought in by the South Sudan civil war, which has contributed directly to the increased commercialization of cattle raiding and child -abduction. The conflict claimed thousands of lives and displacement.

Maban Versus Dinka Adar

The small incident that occurred over the ownership of the Kilo-10 oil-rich area in Northern Upper Nile that lies on the border of Maban and Adar which is claimed by both communities prompted the regime to form a high-level committee in 2017, led by Vice President Wani Igga who has also failed to bring about a tangible solution to the conflict. The conflict has now re-emerged and has taken a tribal dimension that has pitted the people of Maban against the people of Dinka Adar, which has claimed many lives on both sides, and many have been displaced. The regime is unable to bring a lasting solution to the local conflict between the people of Maban and the Dinka Adar oil-rich area.

Cattle Raiding Across Unity and Warrap States Borders

Cattle raiding across the borders of Unity and Warrap states are normally sponsored by the pro-regime militias on both sides of the borders. Pro-regime militias in Unity and Warrap states supply the cattle raiders with ammunition and firearms to raid and get their share of cattle in return. This causes more deaths and encourages more cattle raiding across the border of Warrap and lakes with Unity State. The inter-tribal cattle raiding between communities in Warrap, and Unity State has been there from time immemorial using traditional weapons such as spears, sticks, clubs, and arrows, but this has changed drastically when both communities acquired sophisticated weapons coupled with commercialization, cattle raids, political incitement, and poverty.

Illegal Drugs and the Proliferation of Firearms

The institutionalization of corruption in South Sudan has discouraged and frustrated customs police and other security forces deployed on the borders to thwart the smuggling of illegal drugs and firearms into the country through official border crossing points and porous border areas as that involves high-ranking officials who are collaborating with the drug and firearms dealers who grant them safe crossing passages and trafficking across the borders. The same high-ranking officials are the ones orchestrating economic sabotage, and undermining the market economy, financial system, and nation's resources with the intention to promote impoverished citizens living their lives in a cycle of poverty for a long time. Some senior officials of the regime have also indulged themselves in illegal activities such as contraband goods, money laundering, and other smuggling activities, which are carried out by national economic intelligence through official border crossing points and illegal porous border areas.

Natural Calamities and Ecosystem

South Sudan has been badly affected by floods, especially in some parts of the Western Upper Nile, Western Equatoria, Northern Bhar El Ghazal, and Eastern Upper Nile. The floods have come with water-borne diseases and harmful reptiles. This has caused serious food shortages and a lack of medicines in the affected regions and the regime is unable to provide emergency aid to those areas and has instead asked Egypt to dredge River Naam, which has caused a public outcry on social media. It is not the first time for the people of Western Upper Nile to have experienced the most severe and deadliest flooding over the past centuries.

Floods come because of heavy rains and the construction of more dams on the river Nile that tamper with the natural flow of water. Floods like this occurred severally in the 20th century when there was no government in place in South Sudan, but they managed to survive the floods without being helped by any government.

Now that there is a shaky government in place, it should be its work to keep the people informed about the impending severe weather or give weather updates to the people and have a plan of where to go if there is a flood and which higher ground to be reached quickly on foot or by vehicles. The government should also bear in mind that such floods are inevitable and should have a flood watch mechanism, control methods, and management to reduce or prevent the detrimental effects of flood waters.

The dredging of River Naam is a different scheme of getting money from the Egyptian government and not helping the people suffering. It has nothing to do with the flood and will never prevent or reduce the seasonal flooding in Western Upper Nile. The intention of the Egyptian government is to increase the flow of Nile water so as to have enough water storage for agricultural and irrigation schemes and other long-term plans for water reservoirs. If the government of Egypt was serious to help the people of Western Upper Nile, why

88

do they go direct to River Naam, leaving the flood waters affecting the people outside the river? They should have started to use their machines to make drainage and diversion channels and build dykes and levees in flood-prone areas and move people immediately to the higher grounds.

By dredging River Naam, the machines will tamper with the beautiful river's landscape and ecosystem such as underwater grasses, papyrus, reeds, weeds, etc., which have economic benefits for the local communities and provide natural shelter and food for aquatic wildlife. The plan of the Egyptian government is to remove and glyphosate all the underwater grasses, sudds, papyrus, and reeds, among other plants in and along the bank of River Naam that prevent the normal flow of water into River Nile.

The aim is to dredge River Naam hundreds of meters deep so that all the surrounding swamp fresh waters flow faster into the river, then into River Nile, and within no time there will be no water left in Toch for the livestock except only to get water from River Naam which does not even cover the whole of Western Upper Nile.

If dredging of River Naam is meant for economic and commercial purposes for boats, steamers, and barges to move smoothly from River Nile down to River Naam to Masharar in Tonj, Bhar El Ghazal region like in the past, then the Egyptian government should first destroy the two low-built small bridges on the River Naam in Bentiu and Wangkei respectively. The bridges were built by the Sudan government during the SPLA war for their war equipment to pass through. The bridges were not built for long-term development but can only protect the people from being eaten by the crocodiles and easy crossing for people and livestock.

It would be acceptable if the Egyptian government would construct two high bridges on River Naam to replace the current low-built bridges for boats, barges, and steamers to pass under them to navigate up to Masharar in Warrap State, and they can remove unwanted sudds and other plants in the riverbed that can obstruct the smooth navigation of the barges, steamers, and boats and not to dredge the riverbed or tamper with the river bank landscape. The Egyptian machines can wait until the feasibility study is carried out by River Naam stakeholders and the local communities that comprise people of Guit, Rubkotna, Mayom, Tonj North, Gogrial, Twic, Abiemnom, Abyei, Awiel, and Wau. The government is supposed to consult with the above local communities who are the stakeholders of the Naam water prior to bringing the Egyptian machines to dredge the river.

Environmental Degradation

The environmental degradation in the Republic of South Sudan poses a long-term threat to the ecological system. The forests are degraded and have become sources of income for the suffering soldiers and citizens who have no means of livelihood by resorting to cutting down trees, burning and using them as charcoal for sale in the local markets for their survival. The land, air, and water have been polluted and contaminated, especially in oil-producing areas because of floods, which mix oil with flood water and spill into rivers and other water points where people and animals get drinking water.

The lives of the people in the oil-producing areas are under ecological and physiological threats with babies and animals being born with unknown diseases and deformities. There is no safe and clean drinking water for the local communities and their animals. The regime is supposed to carry out feasibility, physiological and ecological studies for the safety of the communities and their animals in the oil-producing regions and to provide them with health and sanitation facilities and clean water. The government is cheating the people in the oil-producing areas.

The oil-producing states receive five percent as derivation funds monthly, but the impact is not felt in the oil-producing communities. In the last 10 years, the oil-producing states have received millions of dollars from the national government for the oil-producing communities. However, the analysis reveals that intervention activities and physical developments in the oil-producing communities do not reflect the derivation fund received by their states. The derivation fund is paid to the states monthly to assist their oil-producing communities in tackling environmental pollution and degradation, provision basic amenities like healthcare, potable water, paved roads, and economic empowerment of the community people.

Chapter 7

Sudan People's Liberation Army (SPLA)

The national army means the lawful army of the state as distinct from the rebel, paramilitary, and private or militia armies. South Sudan gained its independence in 2011, with the former SPLA rebel army as the nucleus of the would-be professional national army of the Republic of South Sudan. Unfortunately, the country descended into civil war just two years after the country's independence. This meant that the envisaged transformation of the former SPLA rebel army into a professional national army could not take place as the SPLA found itself involved in serious confrontations with rebel groups fighting the regime.

The former SPLA rebel army has been frustrated, weakened, and destroyed by the civil war and has literally been replaced by militia, tribal and factional armies created by both warring parties. There is no national and professional army in the Republic of South Sudan to talk about. The so-called national unified army is still a factional and militia army because they were drawn or drafted from factional or militia army commands and not drawn from the people at the state and county levels. They will remain under their respective political and factional influence for the rest of their service because they owe loyalty only to their respective factional or tribal leaders and not to the nation. The unified factional and militia army has also not been properly trained. They were just given some periodical foot drills that could not qualify them to be the professional national army.

Most of them, if not all deserted the training centres and showed up only at the time of graduation. The absence of a professional national army in the country has facilitated the emergence of tribal and factional militias mainly under various political and tribal leaders. These territorial militias are operating under the pretext of providing security to their cattle and tribesmen and in the process, they are the ones creating rampant insecurity across the country. These militias arm themselves with sophisticated weapons in anticipation of any attack from the other side or rival militias and they are used in the rampant political and inter-communal violence by their tribal and political leaders to pillage villages and commit all forms of atrocities and evil acts.

The SPLA was officially changed from the guerrilla's name to South Sudan People's Defence Forces (SSPDF), on September 2, 2018. The SSPDF, which was supposed to be the nucleus of the formation of the professional national army has become disillusioned with untimely pay, negligence, and lack of welfare. Most of them if not all, have deserted their units and attached themselves to various militia formations in their respective home areas. Furthermore, the so-called unified forces that have recently been passed out from the various training centres across the country, have now begun to follow suit due to a lack of pay, welfare, and negligence.

In 2008, SPLA White Paper was developed to provide a concise overview of its mission and objectives. However, it was not implemented until it was thrown into the dustbin. The white paper codified the military's weakness that requires support, and repair, and created institutions and policy subsets that would require resources and action. The white paper established an SPLA Command Council, reservist force, and Inspector General, in addition to parameters for the Disarmament, Demobilization, and Reintegration (DDR) and civilian conscription and defined the roles of the minister for the SPLA affairs which is for strategic policy and the SPLA Chief of General Staff, which is for operations. The white paper charted out a defensive-oriented SPLA force structure with a sizable peacetime support mandate subordinate to civilian authorities.

In 2005, the SPLA (SSPDF) was officially organized as the nucleus of the professional national army of South Sudan with 10 infantry divisions, eight infantry divisions in South Sudan and two infantry divisions in North Sudan (Nuba Mountain and Blue Nile), before these two former SPLA infantry divisions lost their nationalities when South Sudan became independent in 2011.

The SPLA (SSPDF) current infantry divisions are 1st Infantry Division deployed in Upper Nile with its headquarters in Renk, 2nd Infantry Division deployed in Upper Nile with its headquarters in Malakal, 3rd Infantry Division deployed in Northern Bhar El Ghazal with its headquarters in Wunyik and 4th Infantry Division deployed in Western Upper Nile with its headquarters in Rubkotna-Bentiu.
Others are the 5th Infantry Division deployed in Western Bhar El Ghazal with its headquarters in Wau, the 6th Infantry Division deployed in Western Equatoria with its headquarters in Maridi, the 7th Infantry Division deployed in Eastern Equatoria with its headquarters in Torit, the 8th Infantry Division deployed in Jonglei with its headquarters in Malual Chaat-Bor, 9th Mechanized Division deployed in Jonglei with its headquarters in Ayod, 10th Infantry Presidential Guard Division deployed in Juba with its headquarters in Juba, Special Forces (Commando) deployed at Jebel-Lado and Military Police deployed at the SSDPF headquarters in Juba. The SPLA (SSPDF) is supported by the under-established air force and riverine military branches.

The SPLA (SSPDF) forces are largely officers and Non-commissioned Officers (NCOS) corps with five Major Generals in each infantry division and five Brigadier Generals in each infantry brigade with more over-establishment of other ranks at the lower echelons. The SSPDF and other factional armies in the Republic of South Sudan have lost military standardizations and structure, hence they have become paramilitary in terms of organization and training. Therefore, there is a need for forming a professional and national army in the Republic of South Sudan that shall be drawn from the people at the grassroots level and taken to the national unified training centre for six months training course under the direct supervision of the regional and international military experts. Able-bodied men between the ages of 18 and 45 should be willingly drawn from the following grassroots levels.

1. 30 men from each Boma.
2. 150 men from each Payam.
3. 750 men from each county.
4. 3,750 men from each state.

Further arrangements can be made to phase out the current factional and militia armies along with all the state non-actors and affiliated groups to the regime. This can only be achieved when the UN and other regional bodies are involved to train the new professional national army and arrange for retirement packages for the current factional and militia armies' officers and NCOS corps to phase them out gradually until the new professional national army, the South Sudan Defence Force (SSDF) takes over as the national army of the country and to disarm all the state non-actors or armed civilians in the country and have their weapons destroyed completely to deter cycle of re-armament.

The size of the professional national army shall be determined according to the available external and internal threats in the country but should not be limited to the below-proposed composition of the professional national army of the Republic of South Sudan: -

S/N	Unit	Composition	Deployment	Commanded by
1.	SSDF General HQ	30,000	Juba	General
2.	SSDF Ground Forces HQ	15,000	Juba	Lt. General
3.	SSDF 1st Area Command HQ	3,000	Greater Upper Nile	Lt. General
	1st Infantry Division	10,000	Northern Upper Nile	Major General
	2nd Infantry Division	10,000	Eastern Upper Nile	Major General
	3rd Infantry Division	10,000	Central Upper Nile	Major General
4.	SSDF 2nd Area Command HQ	3,000	Unity, Warrap & Abyei	Lt. General
	4th Infantry Division	10,000	Unity State	Major General
	5th Infantry Division	10,000	Warrap state	Major General
	6th Infantry Division	10,000	Abyei Area	Major General
5.	SSDF 3rd Area Command HQ	3,000	Awiel, Wau & Raja	Lt. General
	7th infantry Division	10,000	Awiel Areas	Major General
	8th Infantry Division	10,000	Wau areas	Major General
	9th infantry Division	10,000	Raja areas	Major General
6.	SSDF 4th area Command HQ	3,000	Greater Equatoria	Lt. General
	10th infantry Division	10,000	Western Equatoria	Major General
	11th Infantry Division	10,000	Central Equatoria	Major General
	12th Infantry Division	10,000	Eastern Equatoria	Major General
7.		G/total	177,000	

Other Details and Composition of the Professional National Army

S/N	Unit	Composition	Commanded by
1.	Field Army	3 Corps = 50,000 soldiers	General
2.	Infantry Corps	3 Divisions = 30-45,000 soldiers	Lt. General
3.	Infantry Division	3 Brigades = 10-15,000 soldiers	Major General
4.	Infantry Brigade	3 Battalions = 3-4,000 soldiers	Brigadier General
5.	Infantry Battalion	3 Companies = 750-1000 soldiers	Colonel
6.	Infantry Company	3 Platoons = 120 -200 soldiers	Major
7.	Infantry Platoon	3 Squads = 30-40 soldiers	2nd Lt or 1st Lt
8.	Infantry Squad	2 teams = 9-11 soldiers	Corporal
9.	Infantry Team	1 Section 4-6 soldiers	Lance Corporal

Now the question is, how do you transform the SSPDF when its nucleus military branches: Land/Ground forces, Air force, Navy, or riverine in the case of South Sudan being landlocked, are commanded by illiterate commanders in this 21st century who do not even know the composition of an infantry Squad leave alone the composition of an infantry platoon and the official paperwork in this modern world full of technology.

The SSPDF has been negligently deprived of military basic welfare benefits. They live only on burning charcoal and fetching firewood from the nearby forest to sell for their survival and their families causing widespread deforestation across the country. The SSPDF is still wearing mismatched uniforms borrowed from other countries like the militia army because the SSPDF command is unable to purchase the SSPDF uniform (**Known as the SPLA Camouflage and Green uniform**) that has been approved by the command council as the country's standard uniform.

The SSPDF has lost discipline, morale, esprit-de-corps, and proficiency because of the lack of timely pay, basic welfare benefits, military equipment, and standard military barracks. The private army of President Salva Kiir along with the National Security forces are well-fed, well-equipped, well-trained, well-uniformed, and well-treated at the expense of the SSPDF. This private army was created in 2012 with the intention to offset the SSPDF which is the nucleus of the would-be professional national army of the Republic of South Sudan.

Territorial Militia Armies in South Sudan

Bul Militia Army, 1985- 2023

Bul-Nuer is one of the seven major Nuer ethnic sections namely Dok, Haak, Nyuong, Jagei, Jikany, and Leek, occupying Western Upper Nile (WUN), from time immemorial, inhabiting west of Western Upper Nile and sharing local and tribal boundaries with Nuer Leek and Jagei in the east, Dinka Twic and Ruweng in the West, Dinka Apuk in the southwest, Dinka Kongor, Lou, and Luacjang in the South and internationally bordering Arab nomads of Misseriya and Nuba in the north, with a roughly estimated population of five hundred thousand inhabitants. Economically, Bul-Nuer lives largely on cattle, fish, and agricultural crops.

Bul-Nuer started to host Militia Army in 1985, when General Paulino Matip Nhial, Nuai-Nuai Nhial, and James Gatduel Gatluak among many other sons of Bul joined the Anyanya II in 1975, at Bilpam on Sudan and Ethiopia border. Paulino Matip and his colleagues abandoned Bilpam when Bilpam came under the SPLA attack in October 1983 and withdrew to South Sudan territory under the overall command of Akuot Atem Mayen, Samuel Gai Tut, and William Abdalla Chuol. Nuai-Nuai Nhial, James Gatduel Gatluak deserted when Bilpam was dispersed, crossed to Western Upper Nile (Bentiu), and joined the Anyanya II group already operating independently in Western Upper Nile without a central command to report to. As a result, Paulino Matip was arrested by William Abdalla Chuol Deng along with other sons of Bentiu David Gatluak Domai and Michael Kel Gatwech. They were accused of sabotage, agitation, and mobilizing the sons of Bentiu to desert with them to the western Upper Nile. Paulino Matip, David Gatluak Domai, and Michael Kel Gatwech managed to escape from custody under William Abdalla Chuol Deng and crossed to the western Upper Nile in January 1984. Paulino Matip met with the SPLA deserters of Tiger and Tumsah Battalions around Fangak on their way to Western Upper Nile. On the way, they nominated Paulino Matip as their commander.

When Paulino Matip arrived in Western Upper Nile in January 1984, he found out that there were guerrilla elements already operating in Western Upper Nile in the name of the Anyanya II under the overall command of Paul Thong Ruay and Robert Ruay Kuol. These were the former Sudan Armed Forces (SAF) soldiers who defected on March 15, 1983, from the Sudan Army garrisons of Wangkei, Mayom, Bentiu, Panaru, and Abiemnom who were part of the deployment of Battalion105, whose headquarters was in Bor under the overall command of Major Kerubino Kuanyin Bol.

It is worth mentioning that, the SPLA mutiny started in Wangkei-Mayom County in Western Upper Nile and not in Bor as claimed by

Bor mutineers, it is only that the headquarters of Battalion 105 was in Bor, but the force deployment was in Western Upper Nile and when the order came to transfer Battalion 105 to northern Sudan, the order was to move first the forces of Battalion 105 deployed in Western Upper Nile areas before the headquarters in Bor could be moved. But as earlier explained, the platoon deployed in Wangkei refused to be transferred to northern Sudan and mutinied by killing their Platoon Commander Second Lieutenant Mohamed Ahmed and crossed river Naam, and stationed themselves in the countryside under command of Corporal James Liah Diu Deng where they were joined later by the rest of their colleagues from Mayom, Abiemnom, Bentiu, and Panaru.

Paulino Matip met with the Western Upper Nile Anyanya-II group and agreed to unify their command under the overall command of Paulino Matip and to be deputized by Paul Thong Ruai. They had promoted themselves to military high ranks in the Anyanya-II command and decided to operate in Western Upper Nile independently in the name of the Anyanya II and carried out some hit-and-run raids on army garrisons in Western Upper Nile including wanton looting and vandalism in oil fields. Major Paul Dor Lampuar, from Leek Nuer, Bentiu was transferred from the SPLA headquarters to Western Upper Nile to go after the SPLA deserters who deserted from the SPLA various divisions that were assigned on different missions from the SPLA central command to regroup, reorganize into the SPLA units and command them in Western Upper Nile zonal command including the Anyanya-II elements already operating in Western Upper Nile without a central command.

Major Paul Dor Lampuar arrived in Western Upper Nile on November 18, 1984, accompanied by the SPLA officers namely: Captain Hakim Gabriel Aloung (his deputy), Second Lieutenant Wilson Deng Kuoirot (Adjutant), and Second Lieutenant Taban Deng Gai (Political commissar), plus other junior officers who were commanding the units. When Paul Thong Ruay, deputy to Paulino Matip heard about the arrival of Major Paul Dor Lampuar, he together with Kot Thian, who were opposing the leadership of Paulino Matip in Western Upper Nile rushed to receive and meet Paul Dor Lampuar in the Jikany area and started to accuse Paulino Matip of supporting the ideology of the Anyanya-II of William Abdalla Chuol Deng in eastern and central Upper Nile.

They asked Paul Dor Lampuar to be their commander instead of Paulino Matip. Major Paul Dor Lampuar came with Paul Thong Ruay and Kot Thian and established his headquarters in Dhorkan and after some time he called Anyanya –II leader Paulino Matip and his group for a meeting at Dhorkan in the Leek area, Rubkona County. When Paulino Matip came with his Anyanya –II officers to the meeting, Paul Dor asked him and his group to abandon the Anyanya-II group's ideology and join the SPLM/A.

Paulino Matip was hesitant at first but was later persuaded by his officers to accept the unification of the forces into the SPLA and to be made deputy to Paul Dor for achieving revolutionary common purposes. Paulino Matip accepted reluctantly in principle to unify the forces under one SPLA command but insisted that he was senior to Paul Dor because he promoted himself as a colonel in the Anyanya-II rank and file and Paul Dor came to Western Upper Nile as the SPLA Major. Major Paul Dor continued to ignore the issues that were raised by Paulino Matip and went ahead and recommended the names of the Anyanya-II officers in Western Upper Nile to the SPLA headquarters for quick integration and confirmation into the SPLA rank and file. The SPLA headquarters quickly confirmed the integration of the Anyanya-II officers into the SPLA with the following ranks: Captain Paulino Matip Nhial, Captain Robert Ruai Kuol, Captain Paul Thong Ruac, Captain James Gatduel Gatluak, Captain Bol Nyawan and Captain Char Makuei Ruea, and other junior officers were made 1st and 2nd lieutenants.

The forces of the Anyanya-II and SPLA in Western Upper Nile were integrated and organized into five SPLA battalions as follows:

1. SPLA Petrol Battalion, led by Captain Paulino Matip Nhial.
2. SPLA Jamus Battalion, led by Captain Bateah Wagah Kueat.
3. SPLA Tiger Battalion, led by Captain Tito Biel Chuor.
4. SPLA Tumsah Battalion, led by Captain Juol Banak.
5. SPLA Koryom Battalion, led by Captain Joseph Mathok Thiep.

Captain Paulino Matip was deployed as commander of the SPLA Petrol Battalion and as well as the deputy to Major Paul Dor in the SPLA western Upper Nile zonal command. The rest of the former Anyanya-II officers were deployed as Company and Platoon Commanders. Captain Paulino Matip was still, however, not happy with the integration of his forces into the SPLA, but Major Paul Dor kept ignoring his complaints. He felt that with the integration, he no longer had direct authority over the forces and civilian population in Western Upper Nile as he used to before the arrival of Major Paul Dor in Western Upper Nile. In his thoughts, he still wanted Major Paul Dor to be his deputy because he was a colonel in the Anyanya-II rank.

Captain Paulino Matip ostensibly asked Major Paul Dor to grant him a few days to go to his home village for some traditional ceremonies, and then come back to take over the command of the new Petrol Battalion assigned to him. Major Paul Dor granted him permission and allowed him to visit his home, which was in the Bul area. Strangely, Captain Paulino Matip informed his loyal soldiers secretly that they should desert after him to the Bul area, and that he was not going to come back to work under Major Pual Dor. He gave them the rendezvous in Bul land to converge but unfortunately, his plan was leaked out before he could leave for Bul area.

On March 10, 1985, a fight broke out between the forces loyal to Major Paul Dor and that of Captain Paulino Matip that were led by Maliny Kawaye Goh on Paulino Matip's side at the parade ground, Nyeromna village. The forces loyal to Paulino Matip were outnumbered and dislodged from the parade ground by the bulk of the SPLA forces that were loyal to Major Paul Dor. Paulino Matip forces were pushed towards Bul area across the Naam River. Many soldiers were killed on both sides and left lying on the parade ground and Paulino Matip crossed the river Naam to Bul land.

On April 15, Paulino Matip Nhial defected from the SPLM/A and declared his allegiance to the Anyanya -II faction led by William Abdalla Chuol Deng with its headquarters and bases in Eastern and Central Upper Nile. The defection of Paulino Matip intensified the fighting between the Anyanya- II forces of Captain Paulino Matip and the SPLA under the overall command of Major Paul Dor Lampuar. The battle was supported by the then Sudan government who started to supply the Anyanya- II forces under Captain Paulino Matip in Western Upper Nile and that of William Abdalla Chuol in Eastern and Central Upper Nile with military supplies to fight against the SPLM/A and called them friendly forces. The Bul area became the battlefield between the SPLA and the Anyanya-II in Western Upper Nile for decades with the destruction of lives and properties. William Abdalla Chuol Deng was killed by the SPLA forces in Fangak area on August 3, 1985, his deputy Gordon Koang Chuol took over the leadership of the Anyanya -II and was deputized by Paulino Matip Nhial from Western Upper Nile Anyanya –II command.

Deployment of Riek Machar to Western Upper Nile

Major Riek Machar Teny-Dhurgon, the SPLA commander of Wolf Battalion (known as Gol Battalion) was deployed in Western Upper Nile by the SPLA headquarters to replace Major Paul Dor and to regroup, reorganize and command the scattered SPLA forces in Western Upper Nile and to make peace with Paulino Matip who

103

had been fighting Paul Dor Lampuar. He arrived in Western Upper Nile on January 1, 1986, to take over from Major Paul Dor but Major Paul Dor refused to hand over the command of Western Upper Nile, claiming that he was not aware of the transfer of Riek Machar to replace him in Western Upper Nile and that he was senior to Major Riek Machar. The disagreement was communicated to the SPLA headquarters by the SPLA military intelligence officer deployed at the headquarters of Major Riek Machar, and it was confirmed that Riek Machar was senior to Major Paul Dor by virtue of being a member of the SPLM/A alternate Political-Military High Command (APMHC) while Major Paul Dor was not a member even though the two were both in the same rank of Major.

A standoff ensued, leading Riek Machar to order the arrest of Paul Dor and send him to the SPLA headquarters under escort. Riek Machar thereafter wrote a letter to Captain Paulino Matip, the leader of the Anyanya-II in Western Upper Nile who had been fighting with Major Paul Dor, asking for reconciliation so that he could join forces with the SPLA to fight the common enemy. Major Riek Machar's letter was misinterpreted to Captain Paulino Matip by his adjutant as if Riek Machar was threatening to fight him when he crossed to Bul land. Paulino Matip got annoyed and instructed his forces to remain vigilant and to be ready to fight Riek Machar whenever he set foot in Bul land. When Riek Machar tried to cross River Naam to Bul area to approach Paulino Matip with the aim of making peace, his forces were attacked by forces loyal to Paulino Matip, an action that led to the outbreak of new clashes between the SPLA under Riek Machar and Paulino Matip's Anyanya -II forces which continued for many years.

On April 16, 1986, Colonel Maliny Kawaye Goh, the strong right-hand man of Paulino Matip was killed in the front-line by the SPLA forces in one of the fierce battles where his head was cut off deliberately and taken to Tharkuer Ciengjoak, the headquarters of Major Riek Machar, and Maliny Kaway's head was hanged and displayed on a pole on the roadside for the public to see. Such barbaric and heinous killing annoyed and instigated Bul to hate Riek Machar and resorted to supporting or standing in solidarity with the Anyanya-II under command of their son Paulino Matip Nhial. Thereafter, the SPLA forces under command of Riek Machar became stronger and more advanced in terms of military equipment and captured May-om on July 7, 1987.

The Anyanya-II forces that were in Mayom withdrew to the Heg-lig oil fields together with the forces of the Sudan Armed Forces (SAF) under command of SAF Captain Ahmed Kabacha. Major Riek Machar Teny-Dhurgon pursued them and attacked the Heglig oilfields but was repulsed with heavy casualties. Mayom was later recaptured the same year by combined forces of the SAF supported by the Anyanya-II militias under the overall command of Brigadier Omer Hassan El Bashir, who was the overall operation commander for the government forces in Western Upper Nile. The SPLA lost more weapons including SPG-9 anti-tank gun which was powerful and new at the time, and as a result, Captain Michael Char Makuei, the SPLA commander of Mayom operations who hailed from Bul Nuer, was immediately arrested by Riek Machar for not putting up a fierce fight to defend Mayom town from falling to the enemy and allowing the SPLA weapons captured by the enemy. While being taken to the SPLA headquarters after being arrested, Captain Char Makuei jumped into River Nile from the boat that was ferrying him across the river and drowned instantly.

Bul accused Riek Machar of killing their son Michael Char Makuei, a claim that was made worse by the killing of Maliny Kaway Goh whose head was hanged on a pole by Riek Machar in 1986. After the capture of Mayom by the SPLA in 1987, Riek Machar ordered the relocation of Bul people from the Bul mainland and settled them across the east bank of Naam river at a place called Kuiynaam or known as Bul-II since he had accused the Bul Nuer people of supporting Anyanya-II faction under command of Captain Paulino Matip. The Bul people were not happy with this move and to make it worse, many of their livestock died from different diseases and people faced famine when they reached the new settlement site. Many people died of hunger and the survivors resorted daily to eating a plant called "Nyakuojuok" which was unfit for human consumption, but Bul ate it for survival, and they named that year 1988, (Ruon-Nyakuojuok) which is loosely translated as the year of famine because they left their houses and everything and lived under hardships without shelter and food to eat.

Paulino Matip established his headquarters in Mayom town alongside the Sudan Armed Forces (SAF) where he received military supplies. He recruited thousands of Bul youth willingly into his militia army to fight against the SPLA. The fighting between the SPLA and the Anyanya II Bul-based militia continued in Western Upper Nile for six years causing wanton destruction of lives and properties on both sides with maximum tribal leverage for Bul to have access to modern and sophisticated weapons that they used to raid and intimidate their neighbours and carried out cattle raids across the border of Bhar El Ghazal with Western Upper Nile.

Gordon Koang Chuol Joined SPLM/A in 1988

The Anyanya-II leader Gordon Koang Chuol and most of the Anyanya -II forces in Eastern Nuer territory negotiated peace with the SPLM/A and joined the SPLM/A in 1988. The two Anyanya-II leaders Stephen Duol Chuol and Gordon Koang Chuol were integrated into the SPLA as full Commanders, and further, Gordon Koang was appointed as an alternate member of the SPLM/A. Captain Paulino Matip promoted himself to the rank of Major General and became the overall leader of the Anyanya -II both in Western Upper Nile and in Eastern Upper Nile, and Abraham Bol Nyathony, who remained after Gordon Koang joined the SPLM/A in Eastern Upper Nile, was appointed deputy to Paulino Matip.

In the same year, 1988, Brigadier James Koang Ruac, the right-hand man of Captain Paulino Matip defected from Paulino Matip's camp in Western Upper Nile to the SPLM/A in Western Upper Nile along with Brigadier Thoare Gatkuer and hundreds of Anyanya-II soldiers, causing a big blow to the leadership of Major General Paulino Matip. Consequently, Brigadier Koang and Thoare Gatkuer were integrated into the SPLM/A as alternate commanders. James Koang Ruac was deployed as the commander of the SPLA in Bul area in Western Upper Nile, and he sent Brigadier Thoare Gatkuer to the SPLA headquarters along with many former Anyanya-II officers mainly from Bul.

Omer Hassan El Bashir in Western Upper Nile

Elder Yap Tekjiek, a Bul Nuer elder gave a traditional spear to Brigadier Omer Hassan El Bashir and blessed him to rule Sudan, El Bashir who was the overall commander of Sudan Armed Forces (SAF) in Mayom area and a long-time friend of Captain Paulino Matip during field operations that led to the recapture of Mayom town from the SPLA in 1987, asked Paulino Matip to provide him with the forces to overthrow the government of Sadiq El Mahdi in the north Sudan. Paulino Matip provided Brigadier Omer Hassan El Bashir with forces mainly from Bul Nuer militias, and he went and took power on June 30, 1989. Bashir pledged to support Paulino Matip with everything he needed to fight on and crush the SPLM/A in a proxy war.

Paulino Matip Nhial worked with President Omer Hassan El Bashir for 17 years as the leader of the Anyanya-II until he joined the government of South Sudan in 2007.

When General Paulino Matip joined the government of South Sudan, another Bul Militia General called Bapiny Manytuil Wicjang, brother to Joseph Nguen Manytuil, the current Governor of Unity State, who formed his own militia army in 2003, the South Sudan Liberation Movement/Army (SSLM/A), and Tut Keaw Gatluak, the crony of Omer Hassan El Bashir, refused to join the government of South Sudan with General Paulino Matip. They decided to remain behind and continue to work with President Bashir.

General Bapiny Manytuil, Tut Keaw Gatluak, and Mathew Puljang Top later joined the government of South Sudan in 2013, which coincided with the SPLM crisis of 2013, and his forces participated in the fight against the SPLA -IO in Unity State and established his headquarters in Bul area, Mayom County under the direct field command of Major General Mathew Puljang Top. The militia forces were deployed in Bul area, Mayom County but were not integrated with the other SPLA forces in 4th Infantry Division in Unity State. However, their leader General Bapiny Manytuil was later integrated into the SPLA as Lieutenant General and assigned as Deputy Chief of General Staff for Moral Orientation.

These forces remain under the indirect command of Bul politicians, especially the Presidential Advisor for Security Affairs, Tut Keaw Gatluak, and Governor of Unity State Joseph Nguen Manytuil. They worked against General Bapiny Manytuil and accused him of rejecting the creation of 28 controversial states that were decreed by President Kiir Mayardit on October 2, 2015. General Bapiny Manytuil fled to Nairobi for his life in 2016, and from there he fled to Khartoum, the capital of Sudan, and formed his own movement under the same name, that is, South Sudan Liberation Movement/Army (SSLM/A).

He came back again after his movement signed the revitalized peace agreement with the government under the umbrella of the South Sudan Opposition Alliance (SSOA). Meanwhile, Tut Keaw Gatluak and Joseph Nguen Manytuil got rid of General Mathew Puljang Top, the commander of General Bapiny's forces in Mayom and dumped him in Juba and replaced him with another commander loyal to them as the commander of Mayom- based militia. These militia forces are currently operating independently in Mayom County under the two politicians, Tut Gatluak Manime and Joseph Nguen Manytuil, Governor of Unity State. They use these militia forces to threaten the President of the Republic of South Sudan against their removal from their current positions. This is the very militia army that carries out cattle raids across the border of Warrap with Mayom and commits atrocities in Leer and Mayendit counties South of Unity state.

Kitgwang and Agwelek Militias

The decision by Riek Machar to return to Juba from rebellion to form a unity government with President Kiir was never well received by the Sudan People's Liberation Movement/Army-in-Opposition (SPLM/A-IO) top commanders.

General Simon Gatwech Dual who was the Chief of General staff (COGS) of the SPLA-IO was against going back to Juba to work with President Kiir in the unified army command without proper security arrangements in which the SPLA-IO and government forces were to be unified to form a professional national army. He wrote to the Intergovernmental Authority on Development (IGAD), the regional block which mediated the peace agreement that ended the war, asking them to release Riek Machar from Juba to enable him to visit and brief his forces on the ground over the implementation of the peace agreement. The IGAD later issued a statement saying there were no restrictions in place against Riek Machar, explaining that Riek Machar was free to visit anywhere of his choice.

A few days later General Simon Gatwech Dual accused Riek Machar of refusing to visit him and decided to suspend the Chief of Intelligence of Riek Machar, Major General Dhiling Keah Chuol who he accused of undermining his authority. Riek Machar refused to endorse the suspension and instead dismissed Simon Gatwech Dual as SPLA-IO Chief of General staff and appointed him as a presidential advisor for peace, a position which Simon Gatwech rejected. Following the development, General Simon Gatwech Dual and

109

General Johnson Olony Thabo called a meeting of SPLA-IO senior officers at Magenis on August 4, 2021, declaring that Riek Machar had been removed as the leader of the SPLM/A-IO and that General Simon Gatwech Dual was the new leader of the SPLM/A -IO. Riek Machar, however, downplayed the move and accused unnamed people in Juba as peace spoilers who had engineered the split within the SPLM/A-IO.

On August 7, 2021, fighting ensued between the forces loyal to Simon Gatwech Dual and Johnson Olony and Tiger Battalion, the headquarters of Riek Machar in Magenis that resulted in the killing of many soldiers on both sides including two senior officers allied to Riek Machar.

The SPLM/A -IO breakaway Kitgwang faction led by Simon Gatwech Dual, and his deputy General Johnson Olony Thabo signed a peace agreement in Khartoum with President Salva Kiir's government. The government bought off General Simon Gatwech Dual and his deputy Johnson Olony Thabo to break away from the SPLM/A-IO and then sign peace with it with an intention to establish militia armies in Lou Nuer and the Shilluk Kingdom to weaken the forces of the SPLM/A-IO to ensure Riek Machar remained in vulnerable and in a weak position and eventually to undermine and sabotage the implementation of the peace agreement.

The agreement was signed by National Security Advisor Tut Keaw Gatluak, Director General of National Security Akol Koor and the SSPDF Chief of Intelligence General Stephen Marshal on behalf of the government, and General Simon Gatwech Dual and General Johnson Olony signed on behalf of Kitgwang faction. General Shamesedeen Al Kabashi of the Sudan Supreme Council signed as a witness, guarantor, and mediator and promised the agreement will end the conflict in South Sudan. Simon Gatwech appreciated President Kiir and promised to return to Juba as soon as possible. The agreement was a military agreement without involving political issues and lacked regional and international back up.

The agreement dealt only with security arrangements including permanent ceasefire, granting amnesty to Kitgwang group and integration of Kitgwang forces into the SSPDF which was agreed to be completed within three months' time after which General Simon Gatwech Dual and Johnson Olony would be in Juba.

Strangely, Johnson Olony was bought off by Tut Gatluak and General Akol Koor who persuaded him to sign a separate peace deal with the government in the name of Agwelek forces, a Shilluk-based militia, so that there would be two peace agreements at the same time. Peace agreement between the government and Kitgwang faction was signed by General Simon Gatwech Dual and his deputy General Johnson Olony, and another peace deal between the government and Agwelek, the Shilluk-based militia was signed by General Johnson Olony as influenced by Tut Gatluak and General Akol Koor.

On February 1, 2022, the Kitgwang faction sent a delegation of 30 officers to Juba to follow up on the implementation of the Khartoum peace agreement that was signed in Khartoum between the government and Kitgwang faction and to pave the way for the coming of Simon Gatwech Dual and Johnson Olony to Juba. The delegation was just dumped in the hotel with limited services until they were thrown out by the owner, citing lack of payment by the government, This prompted Simon Gatwech Dual to write to both the government of South Sudan and Sudan on April 17 and June 14, 2022, demanding that his delegation in Juba return to Khartoum, saying the agreement they signed with the government in Juba was not being implemented. Since the arrival of the Kitgwang delegation in Juba, they had not met any government official or anyone of the parties that they signed the Khartoum peace agreement with. General Simon Gatwech asked the Sudan government who was the guarantor of the Khartoum peace agreement to return their delegation to Khartoum because the mediator (Sudan) had failed to push for the implementation of the security arrangements as agreed in the agreement.

The government reneged on the implementation of the Khartoum agreement with an intention to use General Johnson Olony against General Simon Gatwech Dual. General Johnson Olony accused General Simon Gatwech Dual of removing him from his position as deputy chairman and Commander -in-Chief of Kitgwang faction and appointed Henry Odwar as his new deputy. General Olony argued that when they removed Riek Machar they agreed that they would form a military faction without involving politicians, but now General Simon Gatwech appointed Comrade, Henry Odwar as his deputy without even consulting him.

General Johnson Olony who doubled as deputy Kitgwang faction leader and commander of Agwelek Shilluk-based militia decided to arrest and detain General Simon Gatwech Dual and fly him to Juba as instructed secretly by the Juba regime. General Simon Gatwech Dual discovered the secret of his arrest and escaped safely to Khartoum, Sudan where he took refuge. General Olony rounded up and disarmed Nuer soldiers who he believed were loyal to General Simon Gatwech Dual at Magenis Kitgwang faction's headquarters. However, Nuer soldiers in Tunga area refused to be disarmed and in July 2022, fighting broke out in Tunga between the forces loyal to General Simon Gatwech Dual and General Olony where 18 soldiers were killed on both sides with displacement of 27,000 Shilluk civilians in Panyikang County.

The fighting, which escalated in August 2022, in Diel town, Pigi County triggered further displacement of innocent civilians. The incident that started as a leadership struggle between Generals Simon Gatwech Dual and Johnson Olony Thabo (Kitgwang faction) has now taken an ethnic dimension pitting Nuer of Fangak against the Shilluk tribe in the Shilluk Kingdom. Hundreds of people have been killed and thousands displaced. The conflict has been sponsored on the watch of national security in Juba by political officials on both sides who are supposed to arrest the situation as national-ists. The SPLA-IO in Fangak and Pigi areas has also been dragged into the conflict between Nuer-Fangak and Agwelek-Shilluk-based militias. The fighting between General Simon Gatwech and Agwelek forces escalated when Nuer Gawar of Fangak armed youth joined the conflict, claiming that Agwelek killed their eight fishermen in Tunga prompted the Nuer white army of Fangak to join the fight on the side of General Simon Gatwech Dual.

The combined forces of Simon Gatwech Dual and that of White army supported by the SPLA-IO launched a coordinated attack on Shilluk Kingdom, pushing the Agwelek forces towards the King-dom headquarters in Fashoda where government forces in Malakal crossed to reinforce and save the Kingdom's headquarters. They fought fiercely alongside the Agwelek against the Nuer White army of Fangak and Simon Gatwech Dual forces supported by the SPLA-IO in Fangak and Pigi counties. Thousands of civilians were displaced from Fashoda to Kodok. The mission of Fangak white army that reached up to Fashoda and Wau-Shilluk was led by Fang-ak Spiritual leader Makuac Tut.

Warrap and Awiel Militias

In 2012, the Jieng Council of Elders advised President Kiir to recruit young men from Bhar El Ghazal to protect his leadership in case of any violent coup against him. President Kiir accepted the idea and gave a green light to governors of Warrap Nyandeng Malek and Northern Bhar El Ghazal General Paul Malong Awan who mobi-lized and recruited 10,000 young men from Warrap and Northern Bhar El Ghazal with 80 percent from Awiel and sent them to Pantit training centre where they got trained and named them as Mathiang -Anyor Division in Dinka language which means (Brown caterpil-lar) and the Mathiang-Anyor became the private army to President Kiir. The formation of Mathiang-Anyor was controversial, it was not approved or recognized by the SPLA General headquarters, because it was not in the plan of the SPLA headquarters to make the recruitment across the country which should not be conducted in Bhar El Ghazal alone or from one tribe like Mathiang-Anyor. As a result, the Mathiang-Anyor was stopped by SPLA headquarters from being included in its budget. The Mathiang-Anyor remained without any official budget from the SPLA, but the two governors of Warrap and Awiel took it upon themselves to feed the Mathiang-Anyor from their respective state budgets and sent 3,000 soldiers from Mathiang-Anyor to Juba to increase the strength of the presidential guard division known as Tiger Division where they got their salaries and other services from the presidential budget because it was a private army to the president who was now duty bound to provide for their budget.

In December 2013, Kiir and Riek squabbled in the SPLM about who was going to be the leader of the SPLM, with Kiir accusing Riek Machar of attempting a coup, a move which plunged the country into a civil war which caused a battle between the ethnic Dinka and Nuer in the presidential guard division giving Mathiang-Anyor soldiers the opportunity to do what they were brought to Juba to do ¬– protect the President. The notorious Dinka Maithiang –Anyor and other uncivilized armed Dinka civilians in Juba with influence from some violent Dinka officials and politicians took the law into their own hands, hunted down ethnic Nuer, and targeted them from house to house in the capital Juba, killing over 28,000 from the Nuer ethnic group.

The fighting swiftly spread across the country causing serious destruction in Upper Nile with civilians of either ethnic Dinka or Nuer being targeted on an ethnic basis, launching a full–scale civil war in the country that resulted in the death of nearly 400,000, and displacement of more than two million people. As a result, President Kiir ordered the Chief of General Staff General James Hoth Mai to officially integrate all the Mathiang-Anyor forces into the SPLA to fill the gap of the SPLA soldiers who switched their allegiance to Riek Machar and to deploy the bulk of the Mathiang-Anyor forces to the war front in Upper Nile region, where thousands of Mathiang-Anyor got killed and only a few of them made it to their home villages in Warrap and Awiel. In May 2017, President Kiir fired General Paul Malong Awan as Chief of General staff and replaced him with General James Ajongo Mawut. President Kiir accused Malong Awan of attempting to stage a coup and of power abuse. General Malong became furious and fled Juba towards his home area in Awiel where he wanted to form his rebellion to fight the government of President Kiir, but he was stopped and engaged by local officials and traditional elders in Yirol and eventually he accepted to return to Juba where he was welcomed and received by a cheering crowd at the Juba International Airport on May 13, 2017.

Malong was then put under house arrest by President Kiir but continued to disobey the orders from the President by resisting the arrest which caused tension among his supporters and that of President Kiir because Malong had gained popularity among Ethnic Dinka who saw him as a strongman who saved President Kiir from being toppled by Riek Machar. While under house arrest, Malong's supporters were harassed, detained, and intimidated by Akol Koor who Malong believed was behind his firing as Chief of General Staff until some Dinka elders took it upon themselves to reconcile Malong with President Kiir who accepted to reconcile with Malong and released Malong in November 2017, to go to Nairobi-Kenya for further medical checkup.

In Kenya, Malong announced his defection from the government and formed his own movement called the South Sudan United Front/Army (SSUF/A). When Malong defected, some of the Mathiang-Anyor forces who were mainly made up of Aweil's sons defected and joined Malong's movement which prompted President Kiir not to trust any forces from the Mathiang-Anyor Division. President Kiir ordered for a new mobilization and recruitment to be carried out only from Warrap, his home area, and decided to abandon the Mathiang-Anyor which was mainly made up of Aweil's sons, and called it Malong's army and ordered for the new recruitment to be conducted only in Gogrial, his hometown and about 5,000 young men from Gogrial were mobilized and trained at Panachier village. This division is placed under the command of the Presidential Guard Tiger Division, but secretly deployed at home in Gogrial to offset Malong's supporters in Awiel and Nuer - Bul militias who periodically cross the border and raid cattle in Warrap. One thousand and five hundred soldiers from this force were transferred to Juba to replace the Mathiang-Anyor that was deployed in the presidential guard division in 2012-2013. This division serves two purposes: 1. Protecting Warrap from Nuer - Bul militias and Malong supporters from Awiel, and 2. Part of the SSPDF Presidential Guard (Tiger Division).

Murle - Based Militias

Murle militia has existed since 1980, when Sultan (chief) Ismael Konyi became Murle paramount Chief and Sudan government pro-militia leader in 1987 when the SPLA captured Pibor, and he fled to Malakal where he formed his militia army supported by the government of Sudan with military logistics1. Sultan Ismael was promoted to the rank of Major General and appointed as a militia leader of Murle. Some of Murle's youth joined the Sudan armed forces (SAF) and got promoted to officer rank. Sultan Ismael Konyi was appointed Commissioner of Pibor and worked as Commissioner for 10 years before he was appointed as Governor of Jonglei in 2003, and a member of the National Legislative Assembly in 2005. Sultan Ismael joined the government of South Sudan in 2006, with his militia army when the Comprehensive Peace Agreement (CPA) was signed between the SPLM/A and the government of Sudan in 2005. However, some Murle militias refused to join the government of South Sudan and decided to continue to support the government of Sudan. He was appointed as presidential advisor for peace and reconciliation in 2007, and Governor of Boma State in 2017. Sultan Ismael Konyi was still serving as Sultan of Murle while holding

116

those government positions.

In 2010, David Yau-Yau lost local elections in Jonglei and defected from the government of South Sudan, and formed his own militia movement called South Sudan Democratic Movement (SSDM) with an alternative name known as Cobra faction fighting for the separation of the ethnic minorities of Murle, Anyuak, Kachipo, and Jie from the rest of Jonglei whom he believed being deprived of their rights in South Sudan and to create an independent state. He returned to Juba briefly in 2011 and was promoted to the rank of Major General in the SPLA, but he defected again in 2012.

When President Kiir made an amnesty to all the rebel groups fighting the government, David Yau- Yau declined to respond to the amnesty and continued to fight the government until a peace deal was mediated between Yau-Yau and the government of South Sudan by the church leaders that led to the signing of the final peace agreement in May 2014.

David Yau-Yau was appointed as Governor of Boma State and Chief Administrator of Greater Pibor in 2018-2020 before he was appointed as Deputy Minister of Defence and Veteran Affairs. His militia forces (Cobra faction) were transported to Juba to be integrated into the SPLA, but they all deserted back to Pibor and melted into the civil population and organized cattle raids, child, and women abductions in Jonglei (Greater Bor and Lou Nuer), that forced the SPLA headquarters to create a territorial militia in Pibor to be commanded by the sons of Murle.

These independent territorial Murle militias are the ones responsible for all the atrocities in Jonglei. They call themselves SSPDF while they have not undergone military professional training to be transformed from the militia army to the SSPDF. Former Cobra faction forces partly operate as Murle militias to carry out cattle raids and child abductions in Jonglei and partly as SSPDF forces deployed in Pibor as an independent brigade comprising of one ethnic tribe (Murle).

Maiwut and Jekou-Based Militias

In 2018, the peace agreement that was violated in 2016, as a result of the J-1 battle, was revitalized under the auspices of the IGAD countries, and a peace agreement was signed in Khartoum known as the revitalized agreement on the resolution of the conflict in the Republic of South Sudan (R –ARCSS) that reinstated Riek as 1st vice President and this time around with other four more vice Presidents which resulted in the formation of the fragile and embattled revitalized transitional government of national unity (R-TGONU).

It was agreed controversially to hold elections in December 2022, which failed to take place because the implementation of the peace agreement was deliberately delayed and undermined to buy off Generals of the SPLM/A-IO as in the case of General James Koang Chuol who was politicized and bought off by the SPLM-IG and defected to the SPLM-IG on March 17, 2020, that was followed by the defection of General Simon Gatwech Dual who formed his own Kitgwang militia faction and finally, the defection of the two Gajaak generals, Major General James Ochan Puot and Major General Khor Chuol Giet who defected from the SPLM/A-IO to the SPLM-IG in 2019 and 2021 respectively and established Gajaak-based territorial militia in Maiwut under James Ochan Puot and Jekou under Khor Chuol Giet. Other militias that are sponsored by the regime are community-based militias that protect their cattle and properties from other rival community militias and carry out inter-communal violence across the country. These militia establishments are:

S/N	Territorial militia	Established by
1.	Abyei armed militia	Regime
2.	Arrow boy militia	Regime
3.	Bor Militia	Regime
4.	Koch- based militia	Regime

5.	Maban -based militia	Regime
6.	Mayendit -based militia	Regime
7.	Mayom-based militia	Regime
8.	Torit Manyomeji-based militia	Regime
9.	Renk and Melut-based militia	Regime
10.	Tonj-based militia	Regime
11.	Twic-based militia	Regime
12.	Rumbek and Yirol-based militia	Regime
13.	Mundari-based militia	Regime
14.	Awiel-based militia	Regime

15.	Fangak-based militia	SPLM/A-IO
16.	Lou -based militia	SPLM/A-IO

Note:

Among the many major challenges facing South Sudan, erosion of state authority and growth of ethnic fighting forces stand out as some of the hardest to overcome. The situation is worsened by lack of commitment to transform the SPLA into a professional and disciplined national army capable of supporting peace consolidation. The SPLA (SSPDF) has never been a robust united force since it started to incorporate different militias into it. It is to date still by and large a tool to serve competing interest groups.

The SSPDF has run out of military stockpiles since 2013 and is now only getting its meager supplies and other military equipment on request from the stockpiles of the National Security and Presidential Guard division which is regarded as a total humiliation, subordination, and intimidation.

Chapter 8

Impact of Civil War in South Sudan

South Sudan has hardly ever known peace as conflict continues to torment the oil-rich but deeply poor country years after its leaders declared an end to the civil war estimated to have killed 400,000 people. Despite billions spent by the international community to bring about lasting peace in South Sudan, law and order rarely extend beyond the capital. In February 2023, the UN Mission in South Sudan (UNMISS) warned that armed forces were again mobilizing in Upper Nile State, where artillery and rockets have pounded villages in major offensives involving thousands of troops.

In Jonglei and Greater Pibor, waves of heavily armed youth have carried off women and children in bloody raids targeting their ethnic opponents in recent months. Untold numbers of civilians have died in tit-for-tat massacres in other lawless regions. Tens of thousands have fled to UN bases for protection, compounding what is already Africa's worst refugee crisis. The South Sudanese civil war has created a multitude of crises, including violent clashes between ethnic groups, political instability, food shortages, and many human rights violations.

Attempts by the young nation to hold on to peace snapped in 2013 – within two years of its formation – when an internal political dispute between President Salva Kiir and then vice president Riek Machar exploded into a lengthy conflict that has trapped millions of civilians between government forces and rebel factions.

According to World Bank studies, protracted insecurity and large-scale displacement in the country have taken a huge toll on livelihoods with private consumption consistently falling and the poverty headcount jumped from 51 percent to 82 percent between 2009 and 2016. Conflict-related market closures and disruptions to trade routes have also put pressure on prices.

Humanitarian Crisis

More than 10 years after gaining independence and five years after the signing of the Revitalized Agreement on the Resolution of the Conflict in South Sudan (R-ARCSS), the people of South Sudan continue to face deteriorating humanitarian conditions as a result of conflict, public health challenges, and climatic shocks. Deputy Special Representative in the United Nations Mission in South Sudan (UNMISS) and Resident Coordinator in South Sudan, Sara Beysolow Nyanti says something has to change in the country because the number of people in need continues to rise every year and the resources continue to decrease. Nyanti notes that the deteriorating humanitarian conditions are worsened by endemic violence, conflict, access constraints, operational interference, public health challenges, and climatic shocks such as flooding and localized drought.

United Nations Office for the Coordination of Humanitarian Affairs (OCHA) data shows that an estimated 9.4 million people, a staggering 76 percent of South Sudan's population, will have humanitarian and/or protection needs in 2023. This presents an increase of half a million people compared to 2022. Out of this projected number of needy people in 2023, 1.9 million are internally displaced persons (IDPs), 1.4 million returnees, 5.8 million host community/non-displaced people, and 337,000 refugees. Among them, there are 2.2 million women and 4.9 million children, including 2.4 million girls and 2.5 million boys. Nearly 15 percent of the total people in need are persons with disabilities, according to the UN Office for the Coordination of Humanitarian Affairs.

It is also estimated that eight million people or 64 percent of the population in South Sudan are food insecure and with elevated food insecurity, about 1.4 million children are expected to suffer from life-threatening acute malnutrition. There is also the continued lack of access to safe and improved water and sanitation, especially in areas that are hosting the IDPS. In these areas, women and girls are continuously at risk of being attacked, sexually abused, raped, and harassed by unknown armed men while collecting firewood from the nearby forests or bushes for their families' survival. Some 3.7 million children, adolescents, and caregivers also continue to be at risk of recruitment into local armed groups and other forms of abuse, including abduction and possible trafficking and will need life-saving child protection services in 2023.

South Sudan's health system is among the poorest in the world because there are no basic healthcare services coupled with flooding which is threatening the already fragile health system in the country. Women, children, and the elderly are particularly vulnerable to limited access to health care.

Refugees' Crisis

The protracted political crisis, outbreaks of sub-national inter-communal violence, and natural disasters have colluded to keep millions of South Sudanese out of their homes, with 2.3 million people continuing to live as refugees in neighbouring countries and an additional 2.2 million people internally displaced. According to the United Nations High Commissioner for Refugees (UNHCR), after Syria and Afghanistan, it is the third largest refugee crisis in the world with 63 percent of the refugees are under the age of 18. The majority of those fleeing South Sudan are women and children. They are survivors of violent attacks and sexual assault, and in many cases, children are traveling alone.

Most South Sudanese refugees are living in neighbouring countries such as Sudan, Uganda, Ethiopia, Kenya, and the Democratic Republic of the Congo. According to UNHCR's South Sudan Regional Refugee Response Plan (RRP)1, Sudan is among the largest host countries of South Sudanese refugees, with over 735,000 refugees1 recorded across the country. The Government of Sudan estimates the number of South Sudanese refugees to be over 1.3 million. Despite the closure of the borders due to the Covid-19 pandemic, the government allowed unrestricted access to its territory for over 18,000 South Sudanese refugees who arrived in 2020 through more than 14 different border-crossing points.

Uganda was at the time of the study home to some 889,000 refugees from South Sudan, with over 6,400 new arrivals registered in 2020. Despite Uganda's favourable protection environment, the UN agency has disclosed that refugees are faced with numerous protections challenges due to the magnitude of forced displacement and growing vulnerabilities, compounded by diminishing resources and strained essential social services in refugee-hosting districts. Recent food cuts and Covid-19 measures especially posed additional challenges for refugees in terms of livelihoods and food security. Ethiopia hosted almost 350,000 South Sudanese refugees as of December 31, 2020. Despite the temporary closure of its land borders to prevent the spread of the Covid-19 pandemic, Ethiopia recorded approximately 500 new refugee arrivals from South Sudan in addition to refugees who had spontaneously returned to South Sudan and were subsequently forced to flee again to Ethiopia. Most of the refugees were accommodated in the expanded Nguenyyiel Camp in the Gambella region, where the security situation remained volatile.

In Kenya, most of the 124,000 refugees from South Sudan are hosted in Kakuma camp and Kalobyei settlement in Turkana County. Kenya recorded some 2,250 new arrivals from South Sudan in 2020, the UN explains in the report. The Democratic Republic of the Congo (DRC) hosts some 56,000 South Sudanese refugees. Despite border closures due to the Covid-19 pandemic, 620 new refugee arrivals and 35 South Sudanese refugees who had already been staying for longer periods in the DRC were registered in 2021. The South Sudanese refugee population is staying in a remote part of the DRC, where the security environment is extremely challenging, limiting RRP partners' capacity to reach refugees. The South Sudanese refugee population in the country was projected to increase to around

60,000 by the end of 2022.

Humanitarian Workers

As a result of protracted violence, South Sudan remains one of the most dangerous places for aid workers. According to the United Nations Office for the Coordination of Humanitarian Affairs (OCHA), nine humanitarian workers were killed in the line of duty in 2022 and 450 incidents were reported in the same year while three humanitarian workers have already been killed in 2023. Aid workers are being targeted by unknown armed men as rampant insecurity continues in the country. Humanitarian workers and assets were recently attacked in Pibor when armed youth believed to have crossed from Jonglei broke into an international NGO compound in Pibor with the intention to loot any valuables.

The attack forced World Food Programme (WFP) to temporarily pause its convoy movements out of Bor, Jonglei state, for the second time in as many weeks. Mary-Ellen McGroarty, WFP Country Director in South Sudan described the corridor as critical for its food prepositioning ahead of the rainy season when roads are inaccessible, adding that more than one million people in Jonglei and Pibor rely on the humanitarian food assistance that the UN agency transports along the route. She stressed that the safety and security of staff and contractors are of the utmost importance, adding that when attacks occur, "it is women, men, and children in desperate need of assistance who suffer the most." Between January and December 2021, 591 humanitarian access incidents were recorded. In 2021, five aid workers lost their lives while delivering humanitarian assistance and services.

Humanitarian warehouses and facilities were targeted during the violence, and humanitarian supplies were looted in some locations, significantly impacting response operations in conflict-affected and food-insecure areas. The UN humanitarian agency has called on the authorities to take urgent action to improve security, protect civilians, humanitarian personnel, and commodities, and bring perpetrators to justice.

Economic Impact

South Sudan became the world's newest country and Africa's 54th nation on July 9, 2011. However, outbreaks of civil war in 2013 and 2016 have undermined the post-independence development gains it made, as well as making its economic situation worse. With weak institutions and recurring cycles of violence, South Sudan remains caught in a web of fragility and economic stagnation more than a decade after independence, according to the World Bank. A dearth of economic opportunities and food insecurity remain major concerns and are reinforced by inadequate provision of services, infrastructure deficits, displacement, and recurring climatic shocks. Moreover, poverty is widespread throughout the country because of inter-communal conflict and displacement. The World Bank economic analysis for South Sudan entitled: *Direction for Reform: A Country Economic Memorandum (CEM) for Recovery and Resilience*, highlights the need for the country to leverage its natural capital in the agriculture and oil sectors to support recovery and resilience. Oil and agriculture are the most important sectors of South Sudan's economy, with oil contributing to 90 percent of revenue and almost all exports, while agriculture remains the primary source of livelihood for more than four in five households.

The report suggests that a focus on the country's use of its main endowments of natural capital—oil and arable land—is warranted in the early stages of the country's recovery. The economy of South Sudan in 2023 will still be weak as oil output recovers and public spending continues to rise with rising inflation and food insecurity. The improvement in economic activity will be mainly driven by a rebound in oil exports although domestic demand will weaken as room for public consumption and investment shrinks2. South Sudan is the most oil-dependent economy in the world, with 98 percent of the government's annual operating budget and 80 percent of its gross domestic product (GDP) derived from oil, despite being

125

endowed with adequate natural resources. It has very fertile agricultural land and vast quantities of livestock. The livestock includes over 60 million cattle, sheep, and goats. Instability, unsatisfactory governance, and corruption continue to hinder development in South Sudan. South Sudan is mostly underdeveloped; most cities in the country have no electricity or running water, and overall infrastructure is lacking, with only 10,000 km (6,200 mi) of paved roads. South Sudan is a least developed country according to the United Nations.

1.   South Sudan Regional Refugee Response Plan (RRP).
2.   Fitch Solution Country Risk Research.

Note: *South Sudan remains in a serious economic and humanitarian crisis, and it is only by upholding and fast-tracking the implementation of the peace agreement and strengthening service delivery institutions that can make steps towards becoming the country millions of South Sudanese hoped for when they waved the flag of their new country in the capital Juba on July 9, 2011. Refugees can only come back home and participate in the general elections if there is genuine peace in the country.*

Chapter 9

Land and Boundary Issues in South Sudan

Described to include the surface of the earth and the earth below the surface and all substances other than minerals and petroleum below the surface, land is an essential natural resource, both for the survival and prosperity of humanity, and for the maintenance of all terrestrial ecosystems. In South Sudan, land is a common property of the people which cannot be subjected to sale or any other means of exchange. The customary use of land has been revoked since the country gained its independence on July 9, 2011, and is supposed to be replaced with the national and state laws that shall manage and regulate its use of land.

Land can be divided into urban and communal village land. The portion of land in the urban area is supposed to be allocated to a citizen under an Urban Land Act that shall be made by the legislative branch while the portion of land in the village is supposed to be allocated to a villager or a group of villagers or new settler(s) by the village council under the Village Land Act.

The management of land is supposed to be vested in the national ministry of land and land commission through the national government to regulate the use of land on behalf of the citizens using the law and policy of the land made by the legislative branch. The rights and interests of the citizens in the land shall not be taken without due process of law and a full and fair compensation can be made when land is acquired. Citizens must participate in decision-making on matters related to their occupation or use of land. The national or state governments have a right to seize land of a citizen for public purpose and compensate the owner at fair market value. The national and state governments should regulate and demarcate the land for the following purposes: Residential area, market area, institutional area, industrial and mining areas, road, rail networks and stations, park, reserved land for (wildlife), forest, agricultural area, pastoral area for livestock, recreational areas for (leisure and sports), airports and hazard land (swamps, wetlands, dumping site for hazardous waste, riverbank etc. which should not be developed on account of its fragile nature).

On August 18, 1955, South Sudanese people mutinied in Torit against the Arab regime in the old Sudan and the mutiny spread to all other towns and areas in Southern Sudan. As a result, the other South Sudanese (non-Equatorians) believed that the Equatoria became the source and base of rebellion and willingly started to move to Equatoria region with or without their families, relatives, and friends to join the war of liberation struggle. When peace agreement was signed in 1972, Juba was made the Capital city of Southern Sudan and all non-Equatorians who were in the Anyanya-I around the bushes of greater Equatoria region and those who got assigned in the regional government came to Juba with their families, relatives, and friends, and with time acquired and built portions of land and became citizens in Juba. Indigenous communities in greater Equatoria did not like this development. They began to hate the presence of the non-Equatorians and this bred hatred and hostilities between the non-Equatorians and Equatorian natives in Juba and other towns in greater Equatoria region.

In 1983, Sudan People's Liberation Movement and Sudan People's Liberation Army (SPLM/A) was formed to continue fighting for the war of liberation struggle after the Anyanya movements failed to achieve the independence of South Sudan. The SPLM/A became a formidable and robust resistant movement that deployed thousands of forces in Upper Nile, Bhar El Ghazal and greater Equatoria, capturing towns and establishing bases around greater Equatoria.
In this process, SPLA forces gradually brought their families, relatives, and friends to be with them in their bases or trenches around greater Equatoria region until peace agreement was signed in 2005 between SPLM/A and National Congress Party (NCP), and Juba

was re-confirmed as the Capital city for the Government of Southern Sudan (GOSS). As a result, there was a massive influx of government and military officials who got assigned to the national level and those who came to Juba willingly to look for greener pastures. Juba consequently became a crowded city as everyone sought to acquire a portion of land to build for their families, relatives, and friends. Given non-existence of proper land regulation, these settlements have inflamed serious hatred between non-Bari and Bari indigenous community who call it land grabbing in Juba, a contagious term which has quickly spread to other towns in greater Equatoria region where non-Equatorians, especially Dinka and Nuer are asked to "go back to where they came from."

Dinka Bor Herders in Greater Equatoria

Dinka Bor cattle herders started to move their cattle camps to greater Equatoria region during the SPLM/A-split in 1991, and other subsequent crises in greater Jonglei. They were initially welcomed by the host communities, especially in central and eastern Equatoria, but with time, the local communities began to accuse them of overstaying and destroying their farms and crops and not abiding by the local laws and wanted them to go back to their ancestral land. However, the herders argued that there were still some hostilities in greater Jonglei that could not allow them to go back with their cattle until situation returned to complete normalcy.

Subsequently, the situation has escalated into deadly clashes between the herders and local communities in central and eastern Equatoria causing displacement, death and serious hatred between the cattle herders and the host communities who are influenced by their respective local, state, and national officials. Meanwhile the regime is unable to act because its hands are tied and incapacitated politically and tribally. The hatred and conflict between the cattle herders and local communities escalated into serious clashes in Kajo-Keji on January 24, 2023, where 400 heads of Bor cattle were indiscriminately shot dead in the forest and as a result 20 youth were killed by Bor herders on February 2, 2023.

During the negotiations of the comprehensive peace agreement, especially on oil sharing, Dr John Garang used the concept that the land belongs to the community so that he could take the bigger portion of the oil. However, he did not mean that the land belongs to the community, so the government has no say in it. Of course, the land is owned by the local community, but its management and the policies guiding the ownership lie in the hands of the government of the Republic of South Sudan. It regulates and manages the land on behalf of the community. For if we were to say the land belongs to the community and the government has no say in it, what would happen to oil in Western Upper Nile and Upper Nile (Bentiu and Paloch), which is being managed by the government, in case the local community refuses it to be used or managed by the government, what will happen?

Juba city is now congested and nearly exploding because of lack of space simply because the Bari community does not allow people to be given plots to build on because of the extreme hatred against non-Equatorians who they call land grabbers. Most of the non-Equatorians come to Juba because Juba is the capital city of the country, and not because there is something good in it for the people to grab. Nobody would leave their own birthplace and come to loiter in Juba without doing anything for a living, everybody loves their own birthplace. Bari community wants the capital city to be moved from Juba to elsewhere. Where do they think the capital city should be moved to? Is there any no-man's land? Or it is the land of the people like Bari people? Like Ramciel, is Ramciel a no-man's land? Or is it the land of a certain community? Of course, it is the land of the community and that is why it got the name Ramciel. The idea John Garang had of moving the capital to Ramciel was because Ramciel is situated in the heart of the country. If built, it would be extended to meet with Western Upper Nile in the north, Jonglei in the East, Equatoria in the South and Southwest, Bhar El Ghazal in the West. The same with Juba, if it was allowed to be built, it would be extended to meet with Torit in the East, Yei and Mundiri in the west and Terekeka in the North. Land belongs to the community and must be regulated and managed by the government on behalf of the community.

Issues of Internal Borders

South Sudan's administrative boundaries stemmed from the colonial period. Since it gained independence in 2011, subsequent rounds of reshuffling of the political system, internal borders, and power relations have been a source of confusion, elite manipulation, and conflict throughout the country. The September 2018 Revitalized Agreement on the Resolution of the Conflict in South Sudan commits a Technical Boundary Commission to review administrative and "tribal" boundaries across the country. This comes after the October 2015 announcement of 28 and later 32 states, breaking down the existing 10 states. With May 12 deadline for demarcation decisions, tensions over land and boundaries have become more widespread in South Sudan. The proposed 32 state boundaries may be new, but land disputes in South Sudan are not.

Boundaries have become particularly contentious because people's land rights – and broader citizenship – are increasingly seen to depend on their ancestral belonging to ethnically-defined territories. This reflects governance strategies since the colonial period to fix people in territories and to rule people in groups rather than to recognize individual citizenship rights and freedoms. But it ignores the long history of mobility and fluid social relations among South Sudanese – the idea of exclusive ethnic territory is something relatively new and has been encouraged by recent politics and conflicts. Meanwhile, security improvements following the peace agreement are anticipated to herald the return of South Sudanese refugees and internally displaced people (IDPs), necessitating consideration of support for the returnees' process by the international aid community. After the previous civil wars, the return of IDPs and refugees led to conflicts and tensions over houses, land, and property, but also to debates as to what constituted returnees' areas of origin and rights of residence across South Sudan.

These issues highlight the critical need for reflection about boundaries, citizenship and conflict dynamics surrounding access to and disputes over land in South Sudan. Decentralized governance that has become more prominent in South Sudan since 2005 has also increased the political value of land and contributed to boundary disputes. In South Sudan, territory has become the focus of political and ethnic competition. Control over territory provides access to government revenues and a political constituency. Internal and international boundaries have become more politicized and contested, as neighbouring local administrations seek to maximize their territorial and resource control. One approach to resolving boundary conflicts is to formally demarcate administrative boundaries. This process itself, however, can be a source of conflict.

This is on one hand due to the increasing political and economic stakes in controlling land, and on the other hand because of the non-linear nature of local boundaries to-date. Indigenous boundaries are often indistinct, or interspersed and overlapping, with sometimes only particular clear points such as hills or prominent trees often claimed as boundary markers.

Moreover, historical documentation is only of limited use for demarcating boundaries. There are archival materials on international boundaries, for instance, between South Sudan and Uganda. However, it does not provide the level of detail necessary to present clear solutions to the location of international borderlines. There is even less historical evidence for most other currently disputed internal boundaries. Not only have many of these boundaries recently been created, but previous attempts to define them are vague, inaccurate or unmapped.

Demarcating boundaries is therefore not simply a technical exercise of legally determining and surveying lines, but entails wrestling with the very basis upon which those lines are to be defined—whose claims to land and territory are to be accepted, and with what forms of evidence and what definitions of community. Any process of demarcation requires very sensitive handling of these questions to avoid provoking conflict, as well as substantial support for negotiating local arrangements for cross-border relations, movement, land rights and access to shared resources.

Contestations over administrative territory and boundaries might also play into debates on the return of IDPs and refugees. Attempts to move returnees to their "areas of origin" can be deeply political and relate to tensions over administrative boundaries and ethnic competition over territory. As such, support for such movement's risks entrenching inter-ethnic divisions, which in turn increases the risk of conflict. International aid actors need to consider potential political and conflict dynamics of debates about the return to "areas of origin" and should ensure that they do not facilitate harmfully divisive strategies that foster conflicts and disputes. They must also ensure that returns are not coerced or incentivized, but truly voluntary and based on information about security and services in proposed areas of return. Moreover, returnees might face challenges such as finding out that their land and properties were occupied during their absence. Particularly, in urban and peri-urban areas, land rights were increasingly privatized, individualized and partially commodified through increased formalization of land ownership after 2005.

Yet, land governance, whatever laws, and procedures it uses, tends to be skewed in favour of influential, powerful, and wealthy members of local and national society. In this context, the most constructive responses are those that focus on strengthening the options and institutions for dispute resolution and remedying the grievances of those who are poor, vulnerable, or marginalized in the local politi-

133

cal economy including returnees.

In the new 28 states, President Kiir put all Dinka tribes together as indicated in red on the attached map, which sparked internal border conflicts among the communities, especially in Upper Nile and Western Upper Nile among other areas. In western Upper Nile he put together Dinka Ruweng and Dinka Panaru on the map who have no border between them, since Dinka Ruweng migrated from Panaru to Bul land and were given the portion of land that they are inhabiting now by Bul Paramount Chief Manytuil Wicjang in 1939. Between Dinka Ruweng and Dinka Panaru, are Bul and Leek Nuer who are sharing border with Nuba mountains in the north. Leek Nuer is bordering Jikany Nuer in the East, Dinka Panaru in the far northeast, Nuba Mountain in the north. Bul is sharing border with leek in the east, Twic and Ruweng in the West and Messiriya nomads in the North.

In Upper Nile there is internal border dispute between Shilluk and Dinka Apadang, especially on the western bank of the Nile, Malakal and other surrounding areas which is claimed by Dinka Ngok Lual Yaka and Shilluk. Shilluk claims that Malakal belongs to them, a dispute which has been suspended till further notice.

International Boundaries
South Sudan–Uganda

The modern-day international boundary between South Sudan and Uganda has its origins as a colonial administrative line separating Anglo–Egyptian Sudan from the Uganda Protectorate, both colonies then managed by the United Kingdom. Today, it runs from the tripoint with Kenya in the east for approximately 500 km to the tripoint with the Democratic Republic of the Congo (DRC) in the west. It was initially delimited by a 1914 British Order describing the territory of the Uganda Protectorate. After independence, Uganda unilaterally established its own delimitation in its Constitution, first in 1967, and most recently in 1995. There are several locations on the boundary that were originally described in vague terms, which have led to at least two areas of boundary disputes between South Sudan and Uganda. The first, near the South Sudanese town of Kajo Keji, is an area of contentious sovereignty with a de facto, ill-defined frontier following historic tribal boundaries. For the second, over a larger swath of territory near the South Sudanese town of Parjok, South Sudan and Uganda have explicit opposing claim lines. While there are occasional skirmishes directly related to the unresolved boundary questions, there are much broader problems of instability and conflict, especially in South Sudan due to its civil war. Lack of government control, problems with refugees, and general lawlessness in border regions have made demarcation and formalization of the international boundary between South Sudan and Uganda almost impossible1.

South Sudan–Sudan

After more than a half century of civil war, which began even before Sudan's independence from the United Kingdom, South Sudan became a recognized, separate country in 2011. The international boundary between the two states is based on the provincial boundaries at the time of Sudan's independence from the United Kingdom on January 1, 1956. Much of the modern international boundary remains disputed due to colonial territorial swaps made for administrative convenience and without much heed for the tribal and religious alliances that occurred between the people of northern and southern Sudan. Furthermore, many of the British instruments which defined the boundaries of the various provinces are vague and imprecise, leading to difficulties in establishing a modern delimitation. The boundary separating South Sudan from Sudan extends for about 2,000 kilometres (1,240 miles) from the tripoint with the Central African Republic in the west to the tripoint with Ethiopia in the east. The border region between South Sudan and Sudan remains unstable and prone to violence, with clashes between the militaries of the two governments and between various local communities. Seasonal migration occurs in the border region, which has complicated ideas of land ownership, rights, and sovereignty.

Much of the frontier is considered agriculturally and economically valuable due to the presence of rich farmland and hydrocarbon reserves. Owing to the importance of the frontier area, neither country has been willing to compromise, and the de facto, disputed boundary lingers without much hope of bilateral resolution2. Since the secession of South Sudan from Sudan in July 2011, the two countries are contesting the border areas of Abyei and Mile 14 in Bhar El Ghazal, Kaka El Tijariya, El Fukhar, Magenis mountains in Upper Nile, Heglig in Western Upper Nile and Kafia-Kenygi and Hofra El Nihas in Raja. According to the UN Office for the Coordination of Humanitarian Affairs (OCHA), fighting in the oil-rich border region of Abyei, disputed between Sudan and South Sudan, left 36 people dead as of March 6, 2022, an unknown number injured and 50,000 displaced.

South Sudan–Kenya

The Ilemi Triangle is a disputed territory, which is claimed by Sudan, Kenya, and Ethiopia3. Following numerous efforts to demarcate the area over the last 100 years, Ethiopia, Kenya, and South Sudan have all made conflicting *de jure* and *de facto* claims. This has given rise to a situation in which there is little or no official state involvement in the region and government is at best ad hoc, with all the negative side effects that this entails. The Triangle is home to five major ethnic groups. The nomadic Turkana move in the territory between South Sudan and Kenya and have always been vulnerable to attacks from surrounding communities. The other ethnic groups in this area are the Didinga and Toposa in South Sudan, the Nyangatom, who migrate between Sudan and Ethiopia, and the Dassenach, who live east of the Triangle in Ethiopia. These pastoral people have historically engaged in cattle raiding. Today, pressure on natural resources, principally water and grass, has exacerbated the already tense relations between the ethnic groups. In the past, disputes and conflicts of interest were settled with traditional weapons, today, each group has access to automatic weapons and the loss of life, in even the smallest skirmish, is correspondingly greater. The exact boundaries of the Ilemi Triangle have changed over the years. The British, as the principal colonial power in the region, were instrumental in drawing up the original boundaries.

The Ethiopian Emperor Menelik claimed Lake Turkana and proposed a boundary with the British to run from the southern end of the lake eastward to the Indian Ocean. The British, keen to create a buffer zone between the white settlers in what was then British East Africa and the "wild nomadic tribes of the north" had other plans. The line surveyed by Captain Philip Maud of the Royal Engineers in 1902-3, known as the "Maud Line", put the triangle in Sudan's control and under British hegemony. A subsequent 1907 agreement between Ethiopia and British East Africa was vague on the details of where the border was located. The de-facto border between Kenya and Ethiopia was then set at the Maud line, which ran east-west from the north end of Lake Turkana.

Later in 1914, the Uganda-Sudan Boundary Commission wanted to give Sudan access to Lake Turkana, resulting in the triangular shaped piece of land given to Sudan. Ethiopia, Kenya, and Sudan have each been accused by international observers of using the ethnic groups to fight low-level proxy wars as a means of maintaining their claims to the disputed territory. After the First World War, Ethiopia armed the Nyangatom and the Dassenach, transforming traditional raids into pitched battles in which hundreds of people were killed.

After the 1936 Italian invasion of Ethiopia, a raid against the Turkana in July 1939 by the Italian-backed Nyangatom and the Dassenach led to serious loss of life on both sides. More recently, in the 1980s, it is alleged that the Kenyan government entered into an agreement with the Sudanese People's Liberation Army to administer the contested triangle in return for sanctuary and military and logistical

137

support during the 20-year civil war. Since 1978, the Kenyan government had been suspected of arming the Turkana, while in the 1990s, it was established that Ethiopia supplied the Dassenach with automatic weapons. Now with the discovery of oil, interest in this neglected region has intensified and is likely to increase the possibility of open inter-state conflict. It will come as a further layer of conflict, in addition to the ancient tribal rivalries and the pressure on basic natural resources. Until today, no serious research has been undertaken to study the roots of this multi-layered conflict, and no serious attempts have been made to establish a framework within which a lasting peace could be envisaged. Irish missionary Father Patrick Devine, the founder of the Shalom Centre for Conflict Resolution and Reconciliation, an inter-religious organization, is believed to the first person to try and bring all the principal actors in the region to the negotiating table to examine the issues which have spawned the conflict.

In the recent development, fighting erupted on February 6, 2023, between the Toposa of South Sudan and Turkana of Kenya in Eastern Equatoria State over the common border between the two countries. The clashes occurred after the Turkana community allegedly attacked Toposa in Nadapal, claiming that the pastoralists encroached into their territory. Hundreds of Kapoeta East County residents demonstrated after Kenya deployed soldiers on South Sudanese territory claiming that the border between the two countries is in the Nakodok area of Kapoeta East County. The were later supported by some state parliamentarians who said that the national government's silence on the matter implied that South Sudanese territory had been ceded and or sold to Kenya. South Sudan has accused Kenya of stealing its land, setting the stage for a border dispute that may stymie trade between the two countries. The South Sudanese Minister of Foreign Affairs and International Cooperation, Mayiik Ayii Deng, met with the Kenyan Ambassador to South Sudan, Samuel Nandwa, to discuss areas of mutual concern and later issued a statement assuring all South Sudanese citizens that the highest levels of the government of South Sudan were aware of the sensitivities at the border and were working in cooperation with neighbours to ensure peace, prosperity, and maintenance of border integrity. Kenya and South Sudan share a boundary that stretches for more than 200 kilometres, from the tripoint with Uganda in the south to the tripoint with Ethiopia in the north or east. Over two-thirds of its length is in dispute.

Democratic Republic of Congo–South Sudan

The boundary between Democratic Republic of the Congo (DRC) and South Sudan extends for 813 kilometres and is based on the former colonial border between the Belgian Congo and British Sudan. The border was delimited in the late 19th and early 20th centuries by the colonial powers, and it follows the watershed between the Nile and the Congo Rivers. DRC and South Sudan have not made any adjustments to the border since their independence, and they do not dispute its delimitation.

Central African Republic–South Sudan

The boundary between Central African Republic and South Sudan is based on the colonial border between French Equatorial Africa and British Sudan and was established in 1924. It was delimited and partially demarcated by colonial authorities. Violence and instability in both states has prevented further modern demarcation work from occurring.

The alignment of the Central African Republic–South Sudan frontier is not disputed, but it has a contested region between Sudan and South Sudan, known as Kafia-Kenygi, leaving the precise location of the northern tripoint in dispute and almost 300 kilometres of the eastern boundary of Central African Republic in flux between the administration of Sudan or South Sudan. The established portion of the land boundary extends for an additional 760 kilometres from the tripoint with the Democratic Republic of the Congo in the south to the disputed tripoint with Sudan in the north. It follows the

watershed between the Nile and Congo Rivers for almost its entire extent.
1. Sovereign Limited
2. CSRF-South Sudan, Org
3. Charles Haskins 2009

> Note: *Land in South Sudan continues to be a thorny issue, complicated by the government's land management policy, and the ways these policies conflict with traditional understanding of land acquisition by locals, some who believe their ancestral land cannot be acquired by anybody outside their family lineages as has been seen in greater Equatoria and other regions. These land challenges and occasional border disputes need to be addressed and managed by genuine elected government.*

Chapter 10

Sudan's Endless Crisis

Sudan formally gained independence from Britain and Egypt on January 1, 1956, amid revolt from the southern people. As the country prepared to gain independence, southern leaders accused the new authorities in Khartoum of reneging on promises to create a federal system of government, and of trying to impose an Islamic and Arabic identity on them. In 1955, the Southern army officers were mutinied, sparking off a civil war between the south, led by Anyanya I guerrilla movement, and the Sudanese government.

The conflict only ended when the Addis Ababa Peace Agreement of 1972 recommended the establishment of the Southern Sudan Autonomous Region. The agreement was aimed at addressing the concerns of the Southern Sudan liberation and secession movement. However, in 1983, the South, led by the Sudan People's Liberation Movement (SPLM) and its armed wing, the Sudan People's Liberation Army (SPLA), again rose in rebellion when the Sudanese government cancelled the autonomy arrangements.

The two civil wars waged in South Sudan by the Anyanya I from 1955-1972 and SPLM/A from 1983-2005, eventually gave birth to the independence of the Republic of South Sudan on July 9, 2011. However, other conflicts continued in the western region of Darfur from 2003-2020, displacing two million people and killing more than, 200,000. Since its independence in 1956, Sudan has had more than 15 military coups and ruled by the military junta for most of its existence with only brief periods of democratic civilian parliamentary rule. Omer al-Bashir, a Sudanese military officer who led a revolt that overthrew the elected government of Sudan in 1989 was himself ousted in a military coup in 2019 after weeks of mass protests, with the Transitional Military Council (TMC) taking power.

Lieutenant General Abdel Fattah al-Burhan headed the TMC deputized by General Mohamed Hamdan Dagalo, commander-in-chief
of the Rapid Support Forces (RSF). The protesters were represent-
ed by the Forces of Freedom and Change (FFC) agreed to a power-sharing deal with the military, creating the Sovereignty Council
in August 2019. According to the TMC–FFC agreement, the transition process would last three years and three months. However, on
October 25, 2021, the Sudanese military, led by Gen. Abdel Fattah
al-Burhan, took control of the government in a military coup. He
justified the seizure of power and the dissolution of the authorities
leading the country's transition to democracy, saying infighting between the military and civilian parties had threatened the country's
stability. The RSF, believed to have supported Burhan's rise to Sudan's top job in 2019, was once again heavily involved with the army
in the 2021 coup, which halted the transition to a democratically
elected government, triggering new mass pro-democracy rallies
across Sudan that continue until today.

Rapid Support Force

Rapid Support Force (RSF) is a paramilitary force mainly from the
Darfur region, which was formerly called Janjaweed militias that operated during the Darfur war in the 2000s. President Bashir changed
the name from Janjaweed Militias to Rapid Support Force (RSF) in
2013 and appointed Mohamed Hamdan Daglo (a.k.a Hemeti) as the
commander. Over time, the forces grew and were used as border
guards to clamp down on irregular migration. Analysts estimate the
force to be 100,000, with bases and deployments across the country.

In 2019, former President Omer Hassan El Bashir, called the RSF in
Khartoum to crack down on pro-democracy protesters during the
Khartoum massacre in June 2019. But eventually, with the Sudanese
Armed Forces (SAF), it decided to stand in solidarity with the Suda-
nese Revolution and together they ousted President El Bashir, and
an interim joint civil-military unity government was formed head-

ed by Prime Minister Abdalla Hamdok. Abdel Fattah al-Burhan and Mohamed Hamdan Dagalo, however, seized power in a coup in 2021. This did not change the situation on the ground as after a short period of time tension started building up following intense competition between the army and RSF to recruit new members across Sudan and particularly in Darfur, Hemeti's stronghold. Burhan's proposal to dissolve the Sovereign Council and form a new military council also heightened frictions, as it implied that Burhan could strip Hemeti of his formal political position as deputy chair. After an alarming military build-up in the capital, Burhan and Hemeti reached a deal to de-escalate the situation on March 11, 2023. Hemeti agreed to withdraw forces from greater Khartoum, and the two military leaders agreed to form a new joint security committee. Meanwhile, the other political forces in the country urged the civil population to continue staging protests and demonstrations across the country forcing the junta to sign a formal agreement on April 6, 2023, to hand over authority to a civilian-led government.

But this was delayed because of the latent tensions between Gen. Al-Burhan and Gen. Hemeti, who were serving as chairman and deputy chairman respectively of the transitional sovereignty council. The tense situation was worsened by the dispute over the integration of the RSF into the military, which the paramilitary force resisted, saying they need to operate independently for a period of 10 years before they could be integrated into the Sudan's regular army. However, the Sudanese Armed Forces SAF wants them to be integrated within two years and a decision made on whether RSF should remain under the army chief or under the direct command of Hemeti. The sides led by the two generals have also clashed over the control of the various sectors of Sudan's economy. On April 11, 2023, RSF forces were deployed near the city of Merowe and in Khartoum's key government sites. The SAF issued an ultimatum to the RSF to withdraw their forces from the locations they occupied, but RSF refused. The move made the RSF forces to take control of the Soba military base South of Khartoum.

The RSF began their mobilization, raising fears of potential rebellion which the SAF termed as illegal, and on Saturday, April 15, 2023, the fighting broke out when the paramilitary attacked key government sites in Khartoum by overrunning several key government positions namely the Presidential Palace, Khartoum International Airport, SAF General headquarters, Merowe Airport, Darfur and other states of El Obied, AL Fasher, Genina, Kaduli, Medani, Kosti, Nyala, Kassala, and Gaderif.

The Sudanese Armed Forces responded by using ground forces and military planes to bomb RSF positions, forcing them to withdraw to the nearby neighbourhoods and using civilians as shields. The clashes have, so far, led to the deaths of 400 people, nearly 3,500 injured, and thousands fleeing the capital Khartoum to various states. It also led to foreign governments evacuating their citizens using roads as one side of the airport was under siege by the RSF. The conflict has disrupted humanitarian activities in Sudan where over a third of the population, an estimated 15 million people including the refugees are experiencing acute food insecurity and those trapped inside their homes have no access to water and electricity.

Impact of the Sudan War on the Region

Having been politically unstable for years, the latest outbreak of fighting between forces loyal to the head of the army — Abdel-Fattah Burhan, the country's *de facto* ruler — and his deputy, Mohammed Hamdan Dagalo, who commands the paramilitary RSF group can only imply more instability in the country and a cause of worry for neighbouring countries. Any unrest in Sudan immediately triggers concern in Egypt, Libya, Chad, the Central African Republic, South Sudan, Ethiopia, Eritrea, and — looking across the Red Sea — Saudi Arabia, whether it is for economic, humanitarian, or security reasons.

All of these countries depend on good relations with Sudan, experts point out, especially South Sudan, which despite attaining independence from Sudan in 2011, continues to be beleaguered by Sudan's political and socio-economic ills and insecurity. A shared history still strongly connects people in both countries. Apart from many people still living or staying in each other's countries, including refugees, the two countries enjoy close political and economic ties. Analysts fear that the mass return of the estimated 800,000 refugees living in Sudan could strain the South Sudanese economy already grappling with efforts to supply vital aid to more than two million displaced people who fled their homes because of civil strife.

South Sudan also relies on foreign currency from crude oil sales, which comprises around 95 percent of public revenue and Sudan is crucial to these exports. South Sudan's oil currently runs through Sudan's pipelines to Port Sudan on the Red Sea, where it gets transported to the international market.

Experts also say that foreign interests play a role in the crises and these are likely to shape the way the predicaments in Sudan will take in the future. In the immediate region, Chad and Egypt have maintained close ties with the Sudanese Armed Forces, while the United Arab Emirates and Field Marshal Khalifa Haftar of Libya maintains connections with the RSF. Internationally, the RSF is supported by Russia through its secret company called Wagner, and the SAF is supported by the USA, as the proxy war will continue to rage in Sudan. The mercenary outfit with close ties to the Kremlin, has made inroads across Africa in recent years and has been operating in Sudan since 2017. The United States and the European Union have imposed sanctions on two Wagner-linked gold mining firms in Sudan accused of smuggling.

According to Alex De Waal, a Sudan expert at Tufts University, the fighting in Sudan between the two generals should be seen as "the first round of a civil war." "Unless it is swiftly ended, the conflict will become a multi-level game with regional and some internation-

145

al actors pursuing their interests, using money, arms' supplies and possibly their own troops or proxies," De Waal, who is considered as one of the foremost experts on Sudan and the Horn of Africa, wrote in a memo to colleagues.

South Sudan's Hand in the Sudan War

The Sudan crisis has its root in the South Sudan National Congress party (SSNCP) led by Presidential Security Advisor Tut Gatluak Manime and his team which comprises Dhieu Mathok, Joseph Manytuil Wicjang, the governor of Unity State, Tong Akeen, the governor of Northern Bahr El Ghazal, Dak Duop Bichok and Mayik Ayii. They have hijacked and captured the Presidential Palace of the Republic of South Sudan. This group has controlled the South Sudan presidency and sidelined the real SPLM members from interacting with Salva Kiir, the president. They have managed to manipulate the President into issuing erroneous decrees daily to sack those whom they do not want and appoint whom they want, especially those who can challenge them before the President.

One day, before the war broke out in Sudan on April 15, 2023, Tut Keaw Gatluak, in Khartoum urged Hemeti to take over the power in Sudan, pitting Hemeti against Gen. Burhan, which will pave the way for them to take power in South Sudan. According to a meeting they held on April 3, 2021, they discussed a hint that a doctor had given them regarding the president's ill health. The doctor has hinted that President Kiir will not live for the next six months, and that he will die of liver cancer-related illness. Their discussions, however, revolved on how they would take over when the president dies.

In the meeting, they agreed that the president should remove Nhial Deng Nhial, CDF of the SSPDF, and Gen. Thomas Duoth Guet as external security, then they recommended Marial Benjamin to replace Nhial Deng Nhial, and Gen. Santino Deng Wol to replace Gen. Johnson Juma Okot as a new CDF, and Gen. Simon Yien Makuac to replace Gen. Thomas Duoth Guet in the external security.

According to their scheme of things, when the President passes on, the army and other organised forces will take over the leadership of the country and form a military transitional council (MTC) that will be led by Gen. Akol Koor for two years and be deputized by Security Advisor Tut Gatluak before going for the elections. Gen. Simon Yien Makuac has been brought on board to rally Lou Nuer behind the would-be new leadership. However, if the President does not die as planned, they will topple him in a coup and Hemeti will support them militarily. Hemeti has been briefed about their plan and he has assured them of his greatest support in terms of the military in case the SPLM members try to challenge their action plan.

President Kiir has abandoned his SPLM colleagues in the cold simply because of money supplied by Tut Gatluak. For instance, when Omer Hassan El Bashir was falling from his leadership in Sudan, he sent his wife to Juba for sanctuary with millions of dollars and she was received at the Juba International Airport by Tut Gatluak and Gen. Akol Koor and were residing in the house of Tut Gatluak with the money she brought from Khartoum, but after two days, President Omer was arrested. Then, Tut Gatluak sent the wife of Omer El Bashir to one of the Middle East countries' friends of Omer Hassan El Bashir and the money remained with Tut Gatluak in Juba, promising to send the money after her. But he didn't send the money and instead shared the money with President Kiir, Hemeti, and Gen. Akol Kor.

Secondly, the President also trusts Tut Keaw Gatluak more than any SPLM member because he has promised and assured him that no South Sudanese rebel movements would be harboured in Sudan again to challenge his leadership. Thirdly, President Kiir is surrounded by individuals who appears to scam millions of dollars out of the blue. The money he never had before. Tut Gatluak and his team have also been planning to remove Vice President Taban Deng Gai and replace him with Dak Duop Bichok. The President told them to wait for a while, citing Taban Deng as being a notorious person who can cause havoc in the country. President Kiir has sold the SPLM party and surrendered himself to the South Sudan National Congress Party. Tut Gatluak has already purchased a luxurious real estate for President Kiir and opened bank accounts in one of the Middle East countries where the President intends to retires when his term ends. Tut Gatluak argues that President Kiir should choose the Arab countries as opposed to East Africa reion as his retirement place because the latter will not be conducive for him when he retires, or his property might not be safe.

South Sudan Controversial Map of 28 States showing Dinka territories in Red

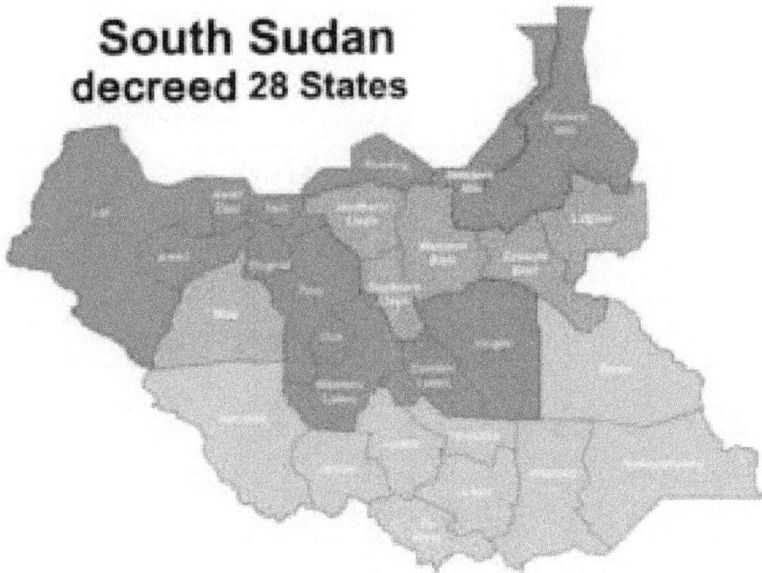

South Sudan Map of 28 States showing other areas

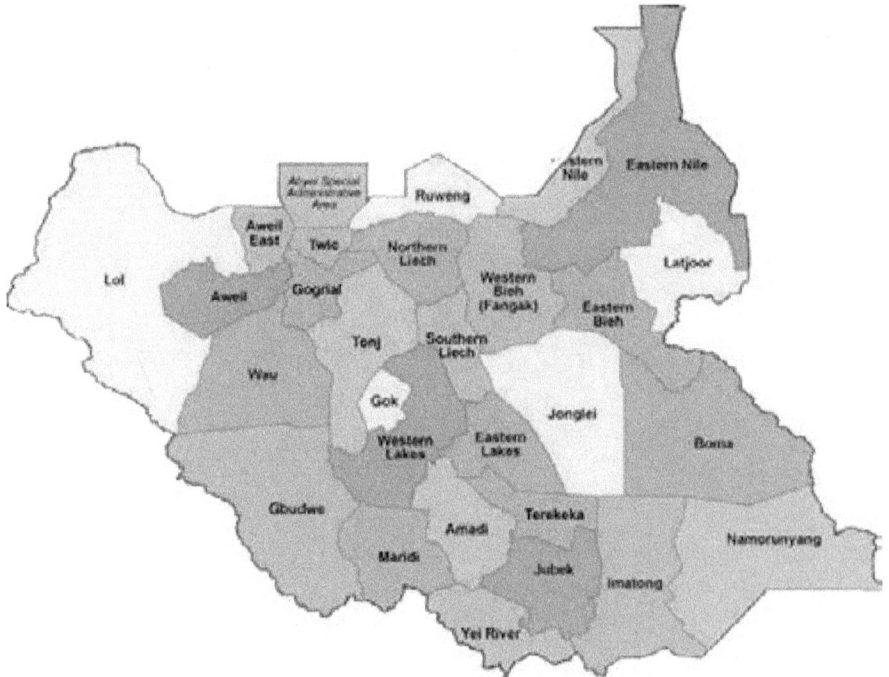

South Sudan Map showing International Boundaries

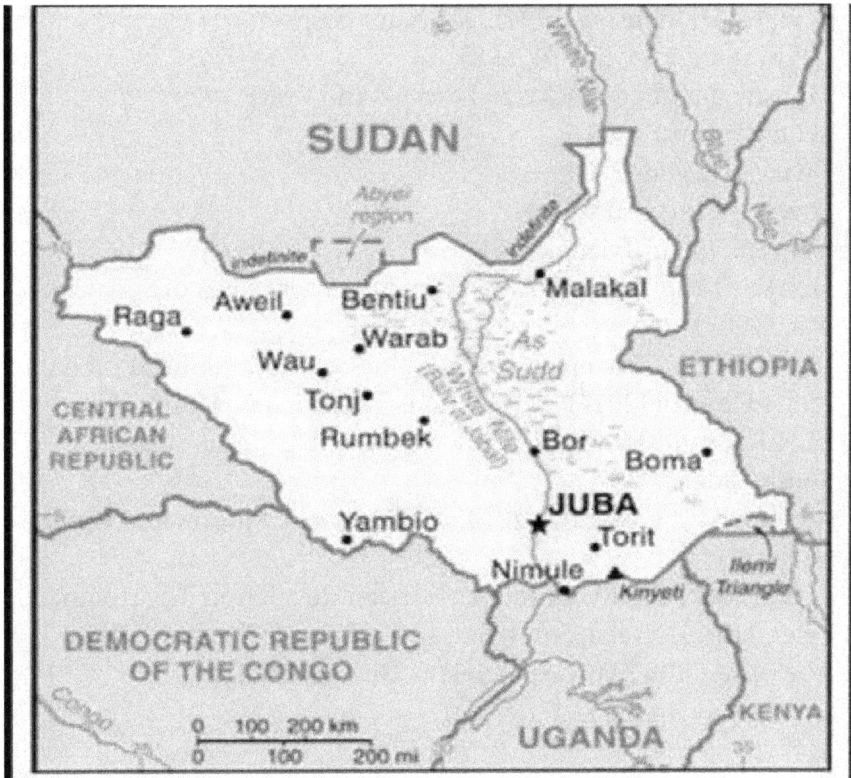

References

1. Douglas H. Johnson (1994): The Nuer prophets
2. SPLM manifesto, July 31, 1983
3. Martin Santschi and Cherry Leonard and CSRF
4. Sentry Report: 2023
5. Charles Haskins 2009
6. Sovereign Limited,
7. Shalom: Illemi Triangle
8. Douglas H. Johnson: African Issues, the root causes of Sudan civil war. (2003)
9. James Bandi Shimanyula (2005): John Garang' and the SPLA
10. James Copnall (2014): A Poisonous Thorn in our Hearts
11. M.H Kanyane, JH Mai, DA Kuok (2009): Liberation struggle in South Sudan
12. Monani Alison Magaya (2014): The Anyanya Movement in South Sudan
13. Lam Akol, SPLM/A: Insider an African Revolution, Khartoum, 2009
14. Peter Adwok Nyaba, South Sudan: Politics of Liberation, 1998,
15. Abel Alier: Too Many Agreements Dishonoured.

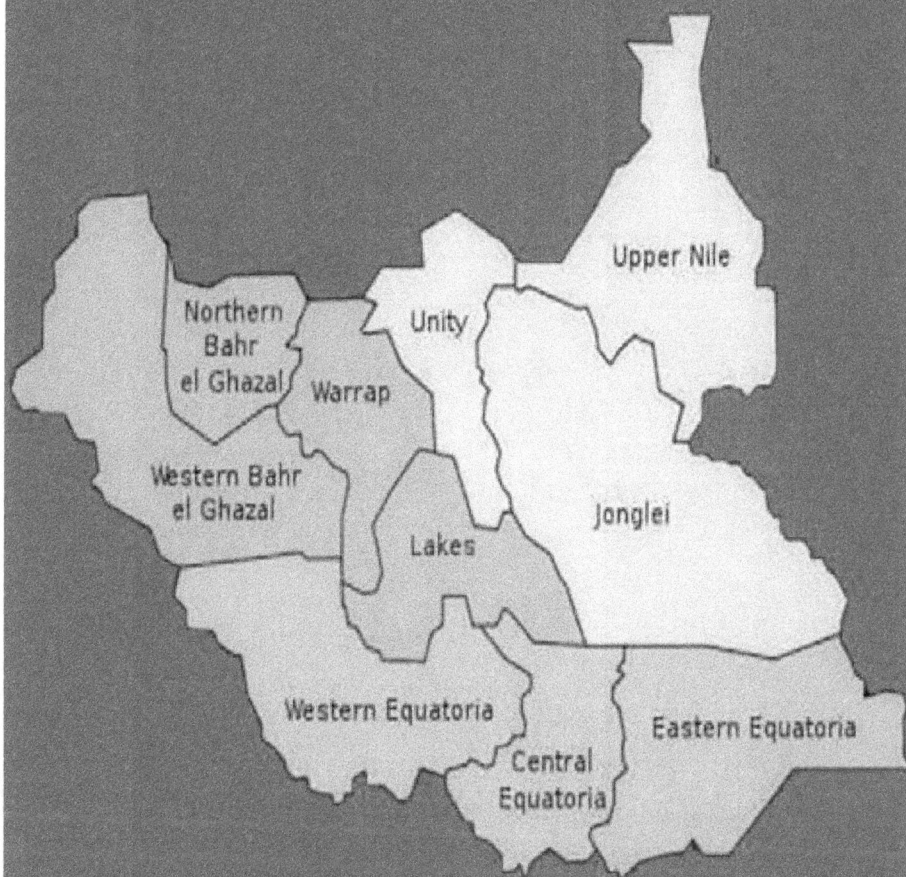

9 780645 819519